Understanding our Quantitative World

© 2005 by
The Mathematical Association of America (Incorporated)
Library of Congress Control Number 2004113543
ISBN 0-88385-738-3
Printed in the United States of America
Current Printing (last digit):
10 9 8 7 6 5 4 3 2 1

Understanding our Quantitative World

Janet Andersen
Hope College

and

Todd Swanson
Hope College

Published and Distributed by
The Mathematical Association of America

BOOK
$58.00

CLASSROOM RESOURCE MATERIALS

Classroom Resource Materials is intended to provide supplementary classroom material for students—laboratory exercises, projects, historical information, textbooks with unusual approaches for presenting mathematical ideas, career information, etc.

101 Careers in Mathematics, 2nd edition edited by Andrew Sterrett

Archimedes: What Did He Do Besides Cry Eureka?, Sherman Stein

Calculus Mysteries and Thrillers, R. Grant Woods

Combinatorics: A Problem Oriented Approach, Daniel A. Marcus

Conjecture and Proof, Miklós Laczkovich

A Course in Mathematical Modeling, Douglas Mooney and Randall Swift

Cryptological Mathematics, Robert Edward Lewand

Elementary Mathematical Models, Dan Kalman

Environmental Mathematics in the Classroom, edited by B. A. Fusaro and P. C. Kenschaft

Essentials of Mathematics, Margie Hale

Exploratory Examples for Real Analysis, Joanne E. Snow and Kirk E. Weller

Geometry From Africa: Mathematical and Educational Explorations, Paulus Gerdes

Identification Numbers and Check Digit Schemes, Joseph Kirtland

Interdisciplinary Lively Application Projects, edited by Chris Arney

Inverse Problems: Activities for Undergraduates, Charles W. Groetsch

Laboratory Experiences in Group Theory, Ellen Maycock Parker

Learn from the Masters, Frank Swetz, John Fauvel, Otto Bekken, Bengt Johansson, and Victor Katz

Mathematical Connections: A Companion for Teachers and Others, Al Cuoco

Mathematical Evolutions, edited by Abe Shenitzer and John Stillwell

Mathematical Modeling in the Environment, Charles Hadlock

Mathematics for Business Decisions Part 1: Probability and Simulation (electronic textbook), Richard B. Thompson and Christopher G. Lamoureux

Mathematics for Business Decisions Part 2: Calculus and Optimization (electronic textbook), Richard B. Thompson and Christopher G. Lamoureux

Ordinary Differential Equations: A Brief Eclectic Tour, David A. Sánchez

Oval Track and Other Permutation Puzzles, John O. Kiltinen

A Primer of Abstract Mathematics, Robert B. Ash

Proofs Without Words, Roger B. Nelsen

Proofs Without Words II, Roger B. Nelsen

A Radical Approach to Real Analysis, David M. Bressoud

She Does Math!, edited by Marla Parker

Solve This: Math Activities for Students and Clubs, James S. Tanton

Student Manual for Mathematics for Business Decisions Part 1: Probability and Simulation, David Williamson, Marilou Mendel, Julie Tarr, and Deborah Yoklic

Student Manual for Mathematics for Business Decisions Part 2: Calculus and Optimization, David Williamson, Marilou Mendel, Julie Tarr, and Deborah Yoklic

Teaching Statistics Using Baseball, Jim Albert

Understanding our Quantitative World, Janet Andersen and Todd Swanson

Writing Projects for Mathematics Courses: Crushed Clowns, Cars, and Coffee to Go, Annalisa Crannell, Gavin LaRose, Thomas Ratliff, Elyn Rykken

MAA Service Center
P.O. Box 91112
Washington, DC 20090-1112
1-800-331-1MAA   FAX: 1-301-206-9789

Preface

Philosophy

Understanding our Quantitative World is our approach to quantitative literacy. This book is intended for a general education mathematics course and addresses the question "What mathematical skills and concepts are useful for informed citizens?" We believe that it is important for students to practice applying mathematical reasoning and concepts to material they are likely to encounter outside academia. Therefore, the text is rich in documented examples taken from sources such as the public media. While we include questions asking students to perform simple calculations, many of the questions focus on using mathematics correctly to interpret information. The topics fall into three categories: interpreting graphs, interpreting simple functions, and interpreting statistical information.

Our goals are for students to:

- Realize that mathematics is a useful tool for interpreting information.

- See mathematics as a way of viewing the world that goes far beyond memorizing formulas.

- Become comfortable using and interpreting mathematics so that they will voluntarily use it as a tool outside of academics.

The text is written in a conversational tone, beginning each section by setting the mathematics within a context and ending each section with an application. Mathematical concepts are explored in multiple representations including verbal, symbolic, graphical, and tabular. The questions at the end of each section are called *Reading Questions* because we expect students to be able to answer most of these after carefully reading the text. Requiring students to read the text before class and to attempt to answer the reading questions allows us to spend class time highlighting key concepts and correcting misconceptions. Having students read the text also emphasizes the importance of becoming a self-learner.

The focus of the course is the *Activities and Class Exercises* found at the end of each chapter. These activities are taken from public sources such as newspapers, magazines, and the Web. Doing these activities demonstrates to students that they can use mathematics as a tool in interpreting the world they encounter. Students spend most of their time in class working in groups on the activities. Rather than having students passively listen, our approach requires students to read, discuss, and apply mathematics.

Students are required to have access to some type of technology such as a graphing calculator or spreadsheet.

The National Science Foundation grant (Grant DUE-9652784) that supported this course also supported the development of two science courses at Hope College, *Populations in a Changing Environment* and *The Atmosphere and Environmental Change*. Connecting this mathematics course with two general education science courses has allowed us to use mathematics as an effective tool in the context of environmental questions and thereby strengthen the students' mathematical understanding.

Annotated Table of Contents

1. Functions. Four representations of functions (symbolic, graphical, tabular, and verbal) are emphasized. Specialized vocabulary (such as domain and range) is introduced. Examples include the stock market, population of the U.S., and the cost of Internet services. Group activities include cell phone rates and credit card bills.

2. Graphical Representations of Functions. Correct interpretation of graphical information is emphasized, particularly with regards to shape and labels. The concepts of increasing/decreasing and concavity are introduced. Instruction on using the calculator to construct graphs is included in the appendix. Group activities focus on analyzing a variety of graphs from magazines, newspapers, and non-mathematical textbooks.

3. Applications of Graphs. The connections between and the meaning of the graphs of $y = f(x)$, $y = f(x + a)$, $y = f(x) + a$, $y = f(ax)$, and $y = af(x)$ are emphasized. This is introduced via the context of a motion detector graph of time versus distance. Group activities include working with a motion detector and converting baby weight charts from English units to metric units.

4. Displaying Data. The emphasis in this section is on visual display of data. Histograms, scatterplots, and xy-line graphs are included. In the appendix, students receive instruction on using the calculator to graph data in each of these formats. Group activities include looking at arm span versus height and data given from the American Film Association on "best movies."

5. Describing Data: Mean, Median, and Standard Deviation. Concepts underlying one variable statistics are emphasized. These include ideas of center (e.g., median and mean) and ideas of spread (e.g. standard deviation and quartiles). The emphasis is on the difference between median and mean, particularly with skewed data. Normal distributions are also introduced. Instruction on using the calculator to compute one-variable statistics is included in the appendix. Group activities include salary versus winning percentage of basketball teams and looking at house prices.

6. Multivariable Functions and Contour Diagrams. Commonly occurring multivariable functions (such as computing the payment on a car loan) and commonly occurring contour maps (such as weather and topological maps) are emphasized. Treating a multivariable function as a single variable function by holding all but one input constant is also included. This allows the students to connect some of the ideas in this section with those encountered earlier in the text. Group activities include a contour map of Mount Rainier and looking at car loans.

7. Linear Functions. The emphasis is on translating a situation with a constant rate of change into the mathematical concept of a line. There is also an emphasis on the concept that only two pieces of information—a starting point and a rate of change—are necessary to determine a line. This section ends by showing that proportional changes (such as unit conversions) can be thought of as linear functions. Group activities include working with a motion detector and looking at an electric bill.

8. Regression and Correlation. Students are introduced to the concept of using linear regression and correlation to determine if two variables exhibit a linear relationship. Calculator instructions for these are included in the appendix. Other types of regression (e.g. exponential) are introduced in later sections. Group activities include Olympic race data and atmospheric carbon dioxide data.

9. Exponential Functions. The concept of an exponential function is introduced via the idea of doubling. Exponential functions are contrasted with linear functions. In particular, the idea of a constant rate of change versus a constant growth factor is emphasized. This section also explores vertical and horizontal shifts of exponential functions, connecting with the ideas introduced in Applications of Graphs. Group activities include a cooling experiment and looking at prices of DVDs.

10. Logarithmic Functions. Logarithms are emphasized as functions that compute the magnitude of a number. Only base 10 logarithms are used. Properties of logarithms and using logarithms to solve simple exponential equations are included. Group activities include working with decibels and verifying Bedford's law on the occurrence of numbers in print.

11. Periodic Functions. Periodic functions are introduced as a way of modeling cyclic behavior. The behavior of a clock and a swing are used to motivate the concepts. Sine and cosine are defined in terms of the circular definitions. The concepts of amplitude and period are related to the ideas of shifting functions introduced earlier in the text. Group activities include an experiment with sound waves and looking at the seasonal change in the amount of daylight per day.

12. Power Functions. Power functions are the last type of function covered in the text and are introduced graphically. Behavior of polynomials with even and odd positive integer exponents is contrasted. Positive rational exponents are also included. Group activities include Kepler's law of planetary motion and looking at the wingspan of birds.

13. Probability. The basic concepts of counting and determining simple probabilities are introduced. Systemic ways of listing (or counting) all possible outcomes are emphasized. Multi-stage experiments and expected value are included. Group activities include codes for garage door openers and roulette.

14. Random Samples. This chapter emphasizes how to set up a random sample and why this is desirable. The concepts of variability, bias, and confidence intervals are included.

Group activities include looking at phone-in surveys and simulating a "capture-recapture" experiment.

Each of these readings is a single unit on the topic. The goal is to give students an intuitive sense of the mathematical concept so they can adequately interpret (rather than necessarily create) mathematics. In addition to the readings, we have also written four to eight group activities for each section, of which we typically assign two to four.

Acknowledgments

We are grateful for the support we have received throughout the project from Hope College. In particular, we are thankful to our colleagues for their encouragement and advice throughout this long process. A special thanks goes to Rolland Swank, Darin Stephenson, Kate Vance, Dyana Harrelson, Mary DeYoung, and Mike Catalano for field-testing our manuscript.

This text was written with support from the National Science Foundation (Grant DUE-9652784). We are thankful for this support.

We are also very thankful to our student assistants, Matt Youngberg, Benjamin Freeburn, Mark Thelen, Melissa Sulok, Sarah Kelly, Todd Timmer, Andrea Spaman, and Kelly Joos for their outstanding work and assistance. Their help has been a valuable part of this project.

We are thankful to the Classroom Resource Materials Editorial Board of the MAA. We are especially thankful to Sheldon Gordon and Zaven Karien for providing their advice and encouragement in the final editing process of our manuscript.

Finally, we wish to thank the staff at the MAA, including Elaine Pedreira and Beverly Ruedi, for their excellent work in producing this book.

Janet Andersen
Todd Swanson

Contents

1

Functions

A function is a mathematical object that uses an input to create an output. It is most commonly thought of as a formula, such as $y = x^2$. Functions can be used to model the world around us. Understanding the behavior of functions allows us to create useful modeling functions for various types of phenomena. In this section, we will give the definition of a function, gain familiarity with the different ways functions can be represented, describe modeling functions, introduce function notation, and describe piecewise functions.

Four Ways to Represent a Function

Functions, in their many representations, are frequently encountered in everyday life. You are probably most accustomed to thinking of functions as symbols, such as the formula for the area of a circle, $A = \pi r^2$. However, you also frequently encounter graphical representations of functions. For example, a graph depicting the behavior of the stock market is a graphical representation of a function. Figure 1.1 shows how the Dow Jones Industrial Average varies over time.[1]

Numerical representations of functions are frequently given as tables of data. For example, Table 1.1 describes the U.S. population as it has changed with respect to time.[2] This table shows us the population at different times to provide a feeling for the trend.

When you are explaining a quantitative situation to someone else, you typically describe the function verbally. For example, the cost of tuition at a midwestern liberal arts college for the 2001–2002 school year is described in the college catalog as follows:[3]

---

[1] *Dow Jones & Company*, "Dow Jones Industrial Average," 21 May 2002, <http://www.djindexes.com/jsp/avgDecades.jsp?decade=2000>.

[2] *U.S. Census Bureau* "Population and Housing Counts," and *American FactFinder*, 2 July 2001, <http://www.census.gov/population/estimates/nation/popclockest.txt>, <http://factfinder.census.gov/servlet/BasicFactsServlet>.

[3] Hope College Official Catalog 2003–2004, p. 79.

Tuition for 12 to 16 credit hours: $9,606.00
Tuition above normal 16-hour load (per credit hour): $215.00
Tuition for 8–11 hour load (per credit hour): $680.00
Tuition for 5–7 hour load (per credit hour): $425.00
Tuition for 1–4 hour load (per credit hour): $290.00

We will use all four representations of functions—symbolic, graphical, numerical, and verbal—throughout this book. All four representations are equally valid. There is not a "right" or "wrong" way to represent a function. Rather, there are four equally valid ways

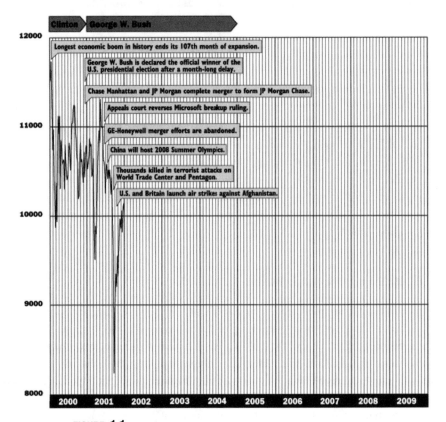

FIGURE 1.1
A graph of the Dow Jones Industrial Average for 2000 to 2009.

TABLE 1.1
The U.S. Population

Year	1900	1910	1920	1930	1940	1950
Population	76,094,000	92,407,000	106,461,000	123,076,741	132,122,466	152,271,417

Year	1960	1970	1980	1990	2000	
Population	180,671,158	205,052,174	227,224,681	249,439,545	281,421,906	

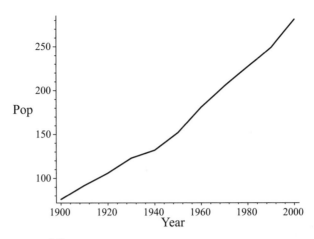

FIGURE 1.2

The U.S. Population in millions of people from 1900 to 2000.

of representing the same thing. However, one of the four representations might work best for a particular situation, depending on what the function represents and what you intend to do with it. Consider the function whose *input* is the year and whose *output* is the U.S. population. To quickly communicate the number of people in a given year, a table of data works best such as Table 1.1. If you want to illustrate how quickly the population is changing, a graphical representation would be better such as Figure 1.2. If you are interested in describing this population function to someone else, you would probably give a verbal description. However, if you want to use this function to predict the population five years from now, you will want to find a symbolic representation. So while all four representations are equally valid, you may prefer one representation over another to convey certain information.

We have explained how to represent a function, but have not yet defined what we mean by the word "function." One of the keys to success in mathematics is understanding the meaning of definitions. Just as with any language, knowing the vocabulary allows you to communicate with others. The definition for function is simple (at least at first glance). A **function** is a "rule" that, for each valid input, assigns one and only one output. Let's look at the four examples given earlier. The formula for the area of a circle is a function since, given a radius, there is one and only one area for the circle. A circle of radius 9 inches has area equal to 81π sq. in. The area cannot be anything else. The graph representing the stock market is a function since, on any given day, there can only be one number that represents the closing Dow Jones Industrial Average. The table of data representing U.S. population is a function since, in a given decade, there is only one official U.S. census count. The description of college tuition is a function since, given the number of credit hours a student takes, there is only one cost for tuition.

The important characteristic of a function is every input has exactly **one** output. This is true whether a function deals with numbers or other objects. For example, the rule:

"Tell me the color of clothing worn by this person."

is not a function since an individual may be wearing more than *one* color of clothing (e.g., the person may have blue pants and a black shirt). However, the rule:

"Tell me the blood type of this person."

is a function since an individual has one and only one blood type. While a function must have only one output for every input, the reverse is not necessarily true. For example, there are many people who have blood type O−. This does not matter when determining whether or not a rule describes a function. What does matter is that every *input* has only one corresponding *output*. The input is often referred to as the independent variable. The output is often referred to as the dependent variable because its value depends on the choice of input.

Determining if a Rule is a Function

No matter how it is represented, the definition of a function is that, for each valid *input*, there is one and only one corresponding *output*. We use this criteria (the definition of a function) to determine if something is a function. This is not necessarily easy. When determining if a rule represented symbolically is a function, you need to reason whether or not any single input can lead to more than one output. There is not an algorithm or procedure that you follow—you just have to logically think about the formula you are given. For example, suppose the relationship between the input, x, and the output, y, is given by

$$y = x^2.$$

This is a function because for any input, x, you have a single output, $y = x^2$, which is the square of your input. There is one and only one number which represents $3^2 = 9$, for example.

However, suppose the relationship between the input, x, and the output, y, is given by

$$y = \pm\sqrt{x}.$$

This is not a function because there is a single input, such as 9, which has more than one output (namely $+3$ and -3).

Typically, it is easy to determine if a rule represented graphically is a function by using the **vertical line test**. If there exists a vertical line that will cross the graph at more than one point, then there is more than one output for a single input so the graph does not represent a function. Figure 1.3(a) is a function since for every input (x-value), there is only one corresponding output (y-value). However, Figure 1.3(b) contains the points $(1, 3)$, $(1, 1)$ and $(1, -2)$. There is a single input, $x = 1$, which has more than one output so this graph is not a function. Notice that, in Figure 1.3(b), some inputs have only one output. For example, the input $x = 0$ has only one output. The important idea when determining if a rule is a function is that *every* valid input has only one corresponding output. If a single input "messes up," it's not a function.

To determine if a table of data represents a function, you need to check whether an input is paired with more than one output. If this occurs, then the table of data does not represent a function. Table 1.2(a) represents a function since each input has exactly one

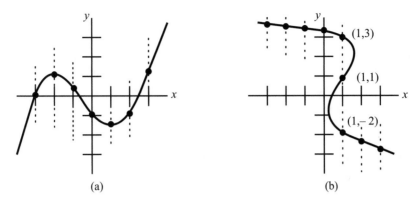

FIGURE **1.3**
The vertical line test. Figure 1.3(a) is a function while Figure 1.3(b) is not.

corresponding output. However, in Table 1.2(b), the input 3 is paired with two different outputs (6 and 8), so this does not represent a function.

TABLE **1.2**
The data in Table 1.2(a) represents a function while the data in Table 1.2(b) does not.

Input	1	3	4	6	9
Output	3	6	8	6	5

(a)

Input	1	3	3	6	9
Output	3	6	8	6	5

(b)

If your rule is represented by a verbal description, it may not be easy to determine if it is a function. As with a symbolic representation, you need to use reasoning to decide if there is more than one output for any single input. For example, the rule:

"Tell me the license plate number and state for this car."

is a function since, for a single car, there will be only one license plate number and state. However, the rule:

"Tell me the e-mail address for this person."

is not a function since a single individual may have more than one e-mail address, i.e., it is likely there will be more than one output.

Modeling Functions

Functions are a useful way of describing the world around us. Most of the functions you encounter (particularly outside of a mathematics classroom) are modeling functions. A **modeling function** is a function that describes a real world phenomenon. The input values and the output values will represent two measurements associated with that phenomenon. Therefore, with a modeling function, the inputs and the outputs will typically have *units* associated with them. For example, $y = x^2$ is a function that gives the square of the input. However, if this function is used to find the area of a square, then $y = x^2$ (or

$A = s^2$) is now a *modeling function* because it represents the area of a square. The input is the length of a side (perhaps given in inches) and the output is the area of the square (then given in square inches).

Modeling functions can be represented in all four ways—symbolic, graphical, numerical, verbal or by combinations of these. All of the examples given in the first part of this reading (area of a circle, stock market, population, and tuition) were modeling functions because they were describing real world phenomena. Notice that the input and output for each of these examples are measurements with specific units such as length in inches or time in years. Which of the four representations you use to describe a particular modeling function depends on what you are describing and how the data associated with that phenomena was obtained. However, there is one situation where a symbolic representation of a modeling function has a clear advantage. Whenever we wish to use a modeling function to predict something that has not yet occurred or to calculate something that is difficult to measure, a symbolic representation is usually the most accurate.

Suppose we wish to use a function to model the population in the United States starting in 1970. The population for 1970, 1980, 1990, and 2000 according to the Census Bureau,[4] is given in Table 1.3.

TABLE 1.3
The U.S. Population

Year	Years since 1970	Population
1970	0	205,052,174
1980	10	227,224,681
1990	20	249,439,545
2000	30	281,421,906
2010	40	????

Representing this function numerically makes it easy to communicate the information. But suppose we want to use this data to predict the population in the year 2010. Just adding the year 2010 to our table and an arbitrary number to the population column is not very effective (and probably wrong)! However, finding a symbolic representation to model this function will allow us to input the year 2010 and use the output as the projected population. The equation of the line that best models our four given data points is:

$$p(t) = 2,513,241t + 203,085,968$$

where $p(t)$ is the population and t is the time in number of years since 1970.[5] This equation allows you to input values of t greater than 30 to estimate the population since

---

[4] *U.S. Census Bureau* "American FactFinder," 30 August 2001,
<http://factfinder.census.gov/servlet/BasicFactsServlet>.
[5] This equation was derived using linear regression—something we will learn about later in this book. The given points will not necessarily lie on the line and, therefore, will not necessarily be solutions to the equation.

the last census. For example, we can predict the population in the year 2010 by inserting the number 40 for t, obtaining:

$$2,513,241 \cdot 40 + 203,085,968 = 303,615,608$$

people. Finding this symbolic representation for our function gives us the ability to predict something which has not yet occurred, assuming that the rate of population growth stays about the same.[6] We can also use this equation to estimate the population for years when the census was not taken. However, remember that our function only provides an estimate of the population assuming no change in things such as the rate of growth of the population. Using it to predict things too far in the future or past will usually give us unreasonable estimates.

Domain and Range

Two important words associated with functions are domain and range. The **domain** of a function is the collection of *valid* inputs. The **range** is the collection of *actual* outputs. For example, in the function:

"Given a person in your class, tell me his or her blood type."

the domain is the set of people in your class and the range is all actual blood types, {A+, A−, B+, B−, AB+, AB−, O+, O−}.[7]

You need to be careful when determining the domain and range of a function. Notice that the domain is limited to the *valid* inputs. When a function is described verbally, or if the function is a modeling function, it is important to consider the context of the function when giving the domain and range. Consider the function $y = \pi x^2$. The domain of this function is all real numbers and the range is $y \geq 0$ (since squaring a number means the output cannot be negative). However, suppose you were told that x represents the radius of a circle and y represents its area. In that case, the domain is $x > 0$ and the range is $y > 0$ since the radius of a circle and its area must always be positive.

Example 1. Find the domain and range of the function:

"Given the length (in inches) of a person's foot, determine his or her shoe size."

Solution. It is clearly inappropriate to say the domain is all real numbers since a foot length of -2 inches or even of 100 inches is *not* a valid size for a person's foot. Instead, the domain is (approximately) 2 inches to 17 inches.[8] The range is limited to actual outputs. It is inappropriate to say the range of our shoe size function consists of all the real numbers because shoe sizes are multiples of $\frac{1}{2}$ between 1 and $29\frac{1}{2}$. ∎

---

[6] Using mathematical models to predict future events can be risky. By making such predictions, you are assuming no changes in the behavior of your system.

[7] The range may actually be a subset of this list since you may not have all blood types represented in your class.

[8] According to the 2002 Guinness Book of World Records, Matthew McGrory has the biggest feet currently known. He wears a size $28\frac{1}{2}$ shoe which corresponds to a foot length of $16\frac{3}{5}$ inches.

Example 2.   We used data giving the weight of infants from zero to six months and found that an equation that relates age to weight is given by

$$w = 1.7m + 7.5$$

where w is the weight in pounds and m is the age in months.[9] Find an appropriate domain and range for this function.

Solution.   It is clearly inappropriate to say the domain is all real numbers since you do not want to use negative numbers to represent age. However, "all positive numbers" is also inappropriate. In fact, if we used the equation for an input such as 36 months, we would find that the predicted weight of an average three year old is $1.7 \times 36 + 7.5 = 68.7$ pounds. But this is twice the weight of a typical three-year old! This is because weight gain is the most rapid during the first few months after an infant is born. Since the equation is based on data giving weight for infants from 0 to 6 months, it makes sense to restrict the domain to $0 \leq m \leq 6$. Using these values as inputs in our function, we see that the range is restricted to $7.5 \leq w \leq 17.7$. ∎

     When a function is represented graphically, the domain is the "shadow" or projection of the graph on the x-axis. The range of your function is the "shadow" or projection of the graph on the y-axis. This is illustrated in Example 3.

Example 3.   Find the domain and range for the function given in Figure 1.4.

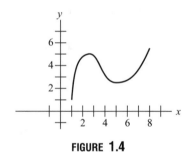

FIGURE **1.4**

Solution.   The domain is the set of all inputs and these are shown on the x-axis, so the domain is $1 \leq x \leq 8$. The range is the set of all outputs and these are shown on the y-axis, so the range is $1 \leq y \leq 6$. These are illustrated in Figure 1.5.

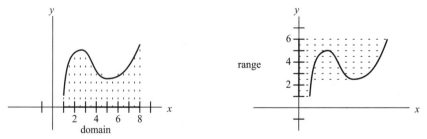

FIGURE **1.5**

---

[9] *THANES*-"National Health and Examination Survey," 12 March 2004, <http://www.cdc.gov/nchs/about/major/nhanes/growthcharts/ charts.htm/>.

When a function is represented numerically, the domain is the collection of numbers in the input (typically first) row or column and the range is all the numbers in the output (typically second) row or column. This is illustrated in Example 4.

Example 4. Find the domain and range of the function given in Table 1.4, assuming the function is limited to the points given in the table.

TABLE **1.4**

Input	1	3	6	9	12	15
Output	3	8	6	5	0	5

Solution. The domain is the set of inputs, or {1, 3, 6, 9, 12, 15}. The range is the set of outputs, or {0, 3, 5, 6, 8}.

When a function is represented symbolically, the two most common things that make an input mathematically invalid are:

- division by zero
- a negative number under a square root sign.

Example 5. Find the domain and range of the function $y = \sqrt{x - 2}$.

Solution. The function $y = \sqrt{x - 2}$ is undefined when there are negative numbers under the square root sign. So the domain of this function is $x - 2 \geq 0$ which implies $x \geq 2$. The range of this function (all actual outputs) is $y \geq 0$ since the outputs will always be nonnegative. ∎

Example 6. Find the domain and range of the function $y = \dfrac{1}{x - 3}$ whose graph is given in Figure 1.6.

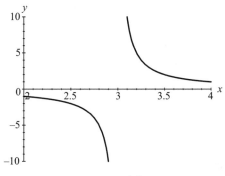

FIGURE **1.6**

Solution. The function $y = \frac{1}{x-3}$ is undefined when the denominator is zero. So the domain is $x - 3 \neq 0$ or $x \neq 3$. The range is all real numbers except zero. It is impossible to divide 1 by a number and get zero for an answer. However, you can get any other number you wish.

Function Notation

The notation $f(x)$ is commonly used to denote a function. This notation, $f(x)$, is pronounced as "f of x." The letter f is typically used to remind us that this object is a *f*unction. The letter x is typically used to denote an input value (i.e., a value in the domain). The f is the *label* representing the function (such as your name is the label representing you) and the x is a *placeholder* representing the input. For example, $f(x) = x + 2$ means "my function, called f, is the rule that adds two to a given input." There is nothing special about the letter x. It should be thought of as a place-holder for your input. So $f(x) = x + 2$ should be thought of as $f(\underline{\quad}) = \underline{\quad} + 2$. For example, $f(5) = 5 + 2 = 7$, $f(-1) = -1 + 2 = 1$, $f(\pi) = \pi + 2$. In fact, $f(y) = y + 2$ and $f(x + 1) = (x + 1) + 2$ because the function f is just the rule "add two to your input".

When a function is a modeling function, it is customary to use letters which represent the physical quantities. For example, the function for the area of a circle is typically written as $A = \pi r^2$ rather than $f(x) = \pi x^2$ to remind us that the input is the *r*adius of the circle and the output is the *A*rea of the circle.

Piecewise Functions

Some functions have different rules for different parts of their domain. For example, the function used for determining income tax is different for different income levels. Such functions are called **piecewise functions**. When determining the output for a given input, it is important to make sure you are using the correct rule. For example, let

$$f(x) = \begin{cases} -x & \text{if } x < 0 \\ x^2 & \text{otherwise.} \end{cases}$$

Then $f(-4) = -(-4) = 4$ since $-4 < 0$, while $f(0) = 0^2 = 0$ and $f(4) = 4^2 = 16$. The decision whether to use the rule "find the opposite of the input" or the rule "square the input" depends on whether your input is less than zero.

Figure 1.7 is the graph of a piecewise function. Notice the abrupt change at $x = 3$. Graphs of piecewise functions often have an abrupt change at the point where the rule changes. The symbolic representation of the function shown in Figure 1.7 is

$$f(x) = \begin{cases} x^2 & \text{if } x < 3 \\ -2x + 15 & \text{otherwise.} \end{cases}$$

Many modeling functions are piecewise functions. For example, America Online,[10] an Internet server, listed their cost as \$9.95 per month for the limited usage plan providing 5 hours and \$2.95 for each additional hour. If we let $c(t)$ represent the cost of using the Internet server for t hours, then $c(t)$ is represented symbolically as

$$c(t) = \begin{cases} 9.95 & \text{if } 0 \le t \le 5 \\ 9.95 + 2.95(t - 5) & \text{if } t > 5. \end{cases}$$

---

[10] *AOL Anywhere* "AOL Pricing Plans," 3 June 2002, <http://www.aol.com/info/pricing.html>.

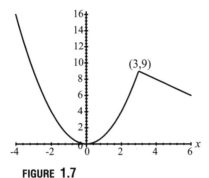

FIGURE 1.7
The graph of a piecewise function.

So the cost of 3 hours of Internet service would be $c(3) = \$9.95$. The cost of 5 hours would be $c(5) = \$9.95$. The cost of 8 hours would be $c(8) = 9.95 + 2.95 \times 3 = \$18.80$.

Application: Internet Costs

Let's see how to use functions to determine the costs of various plans from an Internet service provider. A small Internet service company offers Internet access at the following rates:[11]

Plan	Maximum Number of Hours	Monthly Charge
25 Membership	25	$10
50 Membership	50	$15
Unlimited	unlimited	$18

This is a numerical representation of the Internet information. We can use this information to compare the three plans. For example, we can compare the plans by looking at the cost per hour. To find the cost per hour for the 25 membership plan and the 50 membership plan, assume you will use the maximum number of hours. Then the cost per hour is simply the monthly cost divided by the number of hours. So for the 25 membership plan, you pay \$10/25 hrs or \$0.40 per hour if you use the maximum number of hours. Likewise for the 50 membership plan, you pay \$15/50 hrs or \$0.30 per hour.

Example 7. How many hours would you have to use the unlimited service so that your cost per hour is the same as the cost per hour for the 50 membership plan (assuming you use all 50 hours)?

Solution. Paying \$15 for using 50 hours will have a cost per hour of \$15/50 hrs = \$0.30 per hour. Now we want to know how many hours, at \$0.30 per hour, will give a cost of \$18. To do this, we can write the equation

$$0.30 \times x = 18,$$

[11] *Macatawa Area Community Network Member Pricing and Service List,* 12 March 2004, <http://www.macatawa.org/pricing.html#PersonalMemberships>.

where x is the number of hours. Solving this, we divide both sides by 0.30 to get

$$\frac{0.30x}{0.30} = \frac{18}{0.30}$$
$$x = 60.$$

So you must use the Internet for 60 hours in order for your unlimited membership to give you the same cost per hour as the 50 membership plan. ■

Looking at the calculations in Example 7, we can conclude that if you are not going to use the Internet service for at least 60 hours each month, you are probably better off with the 50 membership plan.

Reading Questions for Functions

1. Name four different ways to represent a function, **giving examples of each**. [Note: The examples must be different from those given in the reading.]

2. (a) Can a function have two outputs for a single input? Why or why not?

   (b) Can a function have two inputs for a single output? Why or why not?

3. Assume a vertical line passes through a graph at more than one point. Explain why the graph does not represent a function.

4. Assume a horizontal line passes through a graph at more than one point. Explain why the graph may represent a function.

5. For each of the following, determine if the rule represents a function. **Write a sentence justifying your answer.**

   (a) Tell me the social security number of a given person.

   (b) Tell me the academic major of a given graduating senior.

   (c)
Input	−2	−1	0	3	2
Output	3	4	6	6	9

   (d) $y = x + 3$

   (e)

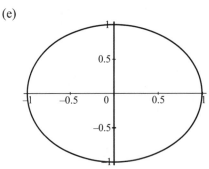

6. For each of the following modeling functions, give reasonable values for the domain and also for the range.

   (a) $V = s^3$ where s is the side of a cube and V is its volume.

   (b) $C = 1.42g$ where C is total cost of gasoline and g is the number of gallons purchased.

   (c) The cost of tuition at your college in a given year.

   (d) The low temperature in your town during the month of December.

7. What is typically the best way to represent a modeling function if you want to use it to predict something that has not yet occurred?

8. Use the symbolic representation of the function that modeled the U.S. population to estimate the U.S. population in 1995. Remember that t is the number of years since 1970.

9. Use the symbolic representation of the function that modeled the U.S. population to estimate the population in 1890. Remember that t is the number of years since 1970. Does your answer seem reasonable? Why or why not?

10. When looking at the following table, what row is associated with the range?

Input	−4	0	3	5
Output	2	4	5	9

11. When finding the domain for a function given symbolically, what two things should you check?

12. (a) What is the domain and range of the function $f(x) = 2\pi x$?

    (b) What is the domain and range of the function $C = 2\pi r$ used for finding the circumference of a circle?

13. Give an appropriate domain and range for each of the following functions. You may use symbols, words, or both when giving your answer.

    (a) $f(x) = x^2 - 2$

    (b) $f(x) = \frac{1}{x-10}$

    (c)

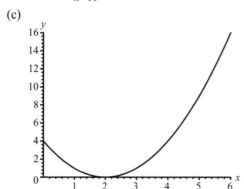

(d)

Input	0	2	4	6
Output	−1	1	2	1

(e) Your summer job pays an hourly wage of $7.25 an hour. Given the number of hours you worked in one week, determine your gross pay for that week.

(f) $F = 5280M$ where M is the number of miles and F is the number of feet.

14. Explain the meaning of the notation:

$$f(2) = 1.$$

15. Let $f(x) = 3x^2 + 4$. Find each of the following:

(a) $f(0)$

(b) $f(2)$

(c) $f(a)$

(d) $f(x + 1)$

16. Let

$$f(x) = \begin{cases} 4x & \text{if } x \leq 0 \\ x^2 - 1 & \text{if } x > 0. \end{cases}$$

Compute each of the following.

(a) $f(-2)$

(b) $f(0)$

(c) $f(3)$

17. Give an example (other than the one in the reading) of a piecewise function which is a **modeling** function.

18. In the reading, we looked at the prices for an Internet service as shown in the following table. We found that the 25 membership plan cost $0.40 per hour while the 50 membership plan costs $0.30 per hour, if you use the maximum number of hours.

Plan	Maximum Number of Hours	Monthly Charge
25 Membership	25	$10
50 Membership	50	$15
Unlimited	unlimited	$18

(a) Why do you think there is a discrepancy in prices per hour for the first two plans?

(b) How much would you charge for the 50-membership plan if you wanted the cost to be $0.40 per hour, assuming all 50 hours were used?

(c) In Example 7, we found the number of hours you would have to use the unlimited membership plan in order for your rate per hour to be the same as the rate per hour for the 50 membership plan. How many hours would you use the unlimited service plan to have the same rate per hour as the 25 membership plan?

Functions: Activities and Class Exercises

1. **Sprint Phone Rates.** Sprint advertised its FONCARD rates as:

   Call from 7 PM to 7 AM each weekday (a total of 12 hours per day) and all weekend long for just a dime a minute. And the rest of the time, 7 AM to 7 PM weekdays, you pay only 40 cents a minute.[12]

   (a) What are the two inputs for this function? What is the output? Include units in your answers.

   (b) Is this a piecewise function? Explain.

   (c) How much does it cost to make a 10 minute call at 8 AM on Wednesday?

   (d) How much does it cost to make a 10 minute call at 8 PM on Wednesday?

   (e) How much does it cost to make a 10 minute call at 6:55 AM on Wednesday?

   (f) Sprint used to advertise itself as the phone company whose rates are just a dime a minute. Explain why this is misleading.

2. **Discover® Card.** Discover® Card is a credit card company that has a *Cashback Bonus Award Program.* Through this program, a cardholder receives a check representing a portion of the amount of money he or she charged each year. The percentage that each cardholder receives depends on the amount of money that was charged according to the following plan.[13]

0.25%	of the first $1000 in purchases
0.50%	of the second $1000 in purchases
0.75%	of the third $1000 in purchases
1.00%	of the amount of purchases in excess of $3,000

   (a) If you charged the following amounts, how much cash would you receive at the end of the year? [Note: For amounts above $1,000, you will have to use more than one rate and add the appropriate amounts.]

   i. $925

   ii. $1,150

   iii. $2,700

   iv. $5,250

   (b) We define the *cashback bonus rate* as the number r such that

---

[12] *Sprint College FONCARD*, 24 May 2002, <http://csg.sprint.com/additional/college_foncard.shtml>.

[13] *Discover® Card Cash Back Bonus Award*, 24 May 2002, <http://www.discovercard.com/discover/data/apply/cashback.shtml>.

$r \times$ amount charged = cashback bonus.

For purchases less than $1,000, the cashback bonus rate is 0.25%. However, for larger amounts, the rate is more complicated since there are different rates for different parts of the domain. If you charged the following amounts, determine the cashback bonus rate, r, by using your answers to part(a) and the formula: $r \times$ amount charged = cashback bonus.

   i. $925

   ii. $1,150

   iii. $2,700

   iv. $5,250

(c) Suppose you heard on a commerial that "You can earn up to 1% of the amount you charge as a cashback bonus award using your Discover® Card." Would you agree or disagree with this statement? Explain.

3. **Tuition Rates.** The information below comes from a 2002–2003 college catalog.[14]

> *Tuition for 12 to 17 credit hours: $7,145*
> *Tuition above normal 17-hour load (per credit hour): $125.00*
> *Tuition for 9 to 11 credit hours (per credit hour): $600*
> *Tuition for 5 to 8 credit hours (per credit hour): $450*
> *Tuition for 1 to 4 credit hours (per credit hour): $300*

(a) Let T be the function that relates credit hours and tuition. Which is the input and which is the output?

(b) Why is this a piecewise function?

(c) How much is the total tuition if you register for 8 credit hours? How much is the total tuition if you register for 9 credit hours? How much more does it cost to register for 9 credit hours rather than 8 credit hours?

(d) If you register for 11 credit hours, how much are you paying per credit hour? How much are you paying for a four-credit class?

(e) If you register for 12 credit hours, how much are you paying per credit hour? How much are you paying for a four-credit class?

(f) If you register for 16 credit hours, how much are you paying per credit hour? How much are you paying for a four-credit class?

(g) If you register for 18 credit hours, how much are you paying per credit hour? How much are you paying for a four-credit class?

4. **Mathematical Predictions.** The following graph is contained in the article "The Coming Job Boom."[15]

---

[14] *Northwestern College*, 4 June 2002, <http://www.nwciowa.edu/view/tuition.html>.
[15] Eisenberg, Daniel. "The Coming Job Boom." *Time* 6 May 2002: 41.

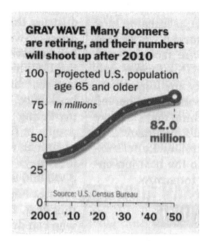

(a) Write a sentence describing what the graph represents. What is the input? What is the output? Include units in your answers. (e.g., If the function was finding the area of a square, your answer should read "the input is length of a side in inches and the output is area of the square in square inches.")

(b) Give a reasonable domain and range for the function represented by the graph.

(c) Notice that the graph includes information about the future. How do you suppose these numbers are determined?

(d) The rate at which the U.S. population age 65 or older is growing is not constant. Why do you think this is?

(e) The header for the graph states that "Many boomers are retiring and their numbers will shoot up after 2010." Yet, what information is actually in the graph? Do you think this graph justifies the header? Why or why not?

5. **Fat Percentages.** When considering the amount of fat in your diet, you need to be aware of not only the total number of grams of fat you are consuming but also the percentage of calorie intake that is from fat. One gram of fat has 9 calories.

(a) The information in Table 1.5[16] is for a single serving. Complete the table by:

   • Calculating the percentage of fat for a single serving in terms of weight (grams).

   • Calculating the percentage of fat for a single serving in terms of calories.

(b) When a label says "x% fat free," is it most likely referring to weight or calories? Why?

(c) Which type of butter has an error in the information on its label? How do you know?

(d) Compare the information for regular butter and lite butter.

---

[16] Data collected at Meijer store in Holland, MI on 24 May 2002.

TABLE **1.5**

Food	Weight (grams)	Calories	Fat (grams)	% of weight due to fat	% of calories due to fat
El Monterey Beef & Bean Burritos (regular)	113	290	14		
El Monterey Beef & Bean Burritos (33% reduced fat)	113	280	9		
Twinkie (regular)	43	150	5		
Twinkie (70% reduced fat)	43	130	1.5		
Land O'Lakes Butter (regular)	14	100	11		
Land O'Lakes Butter (Lite)	14	50	6		
Kraft Ranch Dressing Regular	29	110	11		
Kraft Ranch Dressing (Light Done Right)	31	60	5		

    i. How many grams of non-fat ingredients are in lite butter? in regular butter?

    ii. How many calories are due to the non-fat ingredients in lite butter? in regular butter?

    iii. Based on your answers above, describe how you think the ingredients of lite and regular butter differ.

6. **Cell Phone Rates.** Airtouch offers several different cell phone plans.[17] These include:

Monthly Cost	Free Minutes	Charge per Additional Minute
$35	150	$0.40
$55	400	$0.35
$75	600	$0.35
$100	900	$0.25
$150	1500	$0.25
$200	2000	$0.20
$300	3000	$0.20

---

[17] *Verizon Wireless Prices and Plan*, 24 May 2002, <http://www.verizonwireless.com/ics/plsql/plan/detail.intro?p/hdr/id=25636900&p/plan/category/id=10158&p/section=PLANS/PRICING>.

(a) You decide that you would typically talk on your cell phone for 25 minutes a day. Which plan should you buy? Justify your answer.

(b) You decide that you would typically talk on your cell phone for 60 minutes a day. Which plan should you buy? Justify your answer.

7. **Columbia House.** Columbia has an offer for new members: 12 CDs for a total of $0.01 plus $0.99 each for shipping. However, over the course of two years, you must purchase six additional CDs from Columbia House and pay standard shipping rates of $1.99 per shipment plus $0.99 per CD. You may cancel your membership after you have purchased the additional six compact disks.[18]

(a) What is the initial cost for the 12 CDs and shipping?

(b) What would be the total amount you paid Columbia House if you bought 6 additional CDs at $18.98 each plus shipping and then cancelled your membership? Be sure to include the cost for the initial promotional offer. On average, how much did you pay per CD for the 18 CDs you bought?

(c) What would be the total amount you paid Columbia House if you bought two CDs each month at $18.98 plus shipping for one year? Be sure to include the cost of the initial promotional offer. On average, how much did you pay per CD for the 36 CDs you bought?

(d) If your local record store sold CDs for $14 (tax included), how many CDs would you have to buy from the record store before Columbia House's "twelve CDs for a penny" promotional offer was no longer a better offer?

8. **Cell Phone vs. Phone Card.** Suppose you purchase a cell phone for $149.99 and sign up for a 3500 minutes a month plan which costs $34.99.[19]

(a) Write a function for the cell phone (including the initial cost) when the input is the months you use your cell phone and the output is the total cost.

(b) What is the physical interpretation of the y-intercept?

(c) Suppose you plan on using your cell phone solely for long distance, and you want to decide whether to get a cell phone or a phone card. Assume the phone card costs $16.95 for 500 minutes. We want to determine which would be the more economical buy.[20]

    i. Write a linear equation for both the phone card (assuming you continue to purchase phone cards) and the cell phone where the input is **hours** used and the output is the cost.

    ii. Graph the two functions and determine from the intersection how many hours it would take for both options to cost the same.

---

[18] *"ColumbiaHouse.com Offer Detail Page"*, 5 June 2002, <http://www.columbiahouse.com/cl1/ch/offer/get_offer_details.jsp?&club=1&stype=JOIN&dt=1023304876650&pin=72&bak=71>.

[19] *Sprint PCS—The clear alternative to cellular*, 9 July 2002, <http://www.sprintpcs.com>.

[20] *Phone cards On sale*, 9 July 2002, <http://www.phonecardonsale.com/index.html?action=1&todo=center_usa.html&sid=3D2AFBDCC66E6A970702>.

iii. If you were to use all 3500 minutes a month for your cell phone plan, on average how many hours a day would you have to be talking on your phone? Base your answer on a thirty day month.

iv. Suppose you were considering the phone card because you estimate you will only spend around 20 hours each month on the phone. Write the linear equation for the phone card with months for input and cost as the output. Graph this equation along with the one you found in part (a). How many months will it take for the cell phone to become more economical than the phone card?

2

Graphical Representations of Functions

You have probably heard the saying, "A picture is worth a thousand words." In many ways, this is true. You can read about the Grand Canyon, the New York skyline, or Mount Rainier, but words do not give you the same understanding as just looking at a picture. The same principle is true of the relationship between graphic and symbolic representations of a function. Looking at the graph of a function often gives you a better understanding than looking at the symbols. In this section, we will look at the importance of graphs, some common vocabulary used to describe them, and how to construct them using a calculator.

Constructing a Graph

The types of graphs we will look at include contour graphs, bar graphs and line graphs. Newspapers and magazines publish line graphs to track the stock market and contour graphs to show weather patterns. Since we frequently encounter information graphically, knowing how to interpret graphs appropriately is a valuable skill. In this section, we will focus our attention on graphs of functions plotted in the Cartesian,[1] or rectangular, coordinate system. Other types of graphs will be discussed in later sections. The Cartesian coordinate system is shown in Figure 2.1.

The graph in Figure 2.1 consists of two axes—a horizontal axis (often labeled the x-axis) and a vertical axis (often labeled the y-axis). The position of a point on the graph is denoted by an ordered pair such as $(2, 4)$. An ordered pair consists of two numbers where the first number refers to the horizontal position on the x-axis and the second number refers to the vertical position on the y-axis. The number for the horizontal position typically represents an input for the function while the vertical position number typically represents an output. The two axes intersect at a point called the **origin** which

---

[1] The Cartesian coordinate system is named after Rene Descartes who first used the two-axis system to plot points. He is the same person who philosophized about his existence by saying "I think, therefore I am."

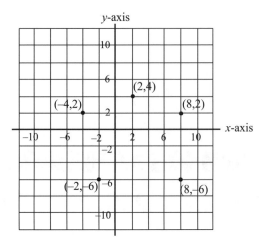

FIGURE 2.1
Cartesian graph with several points labeled.

has the coordinates $(0, 0)$. A negative horizontal number means you move the specified number of units to the *left* of the origin. A negative vertical number means you move the specified number of units *down* from the origin.

The chapter on *Functions* mentioned four ways of representing functions (symbolic, numerical, graphical, and verbal). One advantage of describing data in graphical form is that it allows you to observe trends and see overall behavior quickly and easily. We will illustrate this using outside temperature data collected on a day in July with a Calculator Based Laboratory (often refered to as a CBL) and a TI-83 calculator. The program first stored the data in the calculator numerically and then graphed it. Some of these data are given in Table 2.1.[2] Looking at the data gives a lot of information, but it takes some time to see relationships or trends.

However, a graph of this data (shown in Figure 2.2) allows you to quickly and easily see the relationship between time and temperature. To make this graph, we plotted the points and then connected them with short line segments. Notice that the temperature initially

TABLE 2.1
Temperature data (in degrees Fahrenheit) for a midwestern town on a day in July. (Note: Not all of the collected points are shown.)

Time	4 PM	5 PM	6 PM	7 PM	8 PM	9 PM	10 PM	11 PM	12 AM
Temp.	72	69	65	63	62	62	61	59	58
Time	4 AM	5 AM	6 AM	7 AM	8 AM	9 AM	10 AM	11 AM	12 PM
Temp.	57	62	66	72	76	80	81	82	84

---

[2] These temperatures were collected in Holland, Michigan using a TI-83.

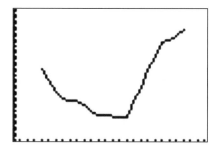

FIGURE 2.2
A calculator graph of the relationship between time and temperature for the data set given in
Table 2.1.

decreased, leveled off for a short time, then increased fairly rapidly before starting to level
off again towards the end.

The graph in Figure 2.2 is not as useful at it could be. Since this graph has no labels
or tick marks on the axes, the information it provides is limited. Without labels, the only
information we can obtain from a graph is its shape. Looking at Figure 2.2, you can
see that the temperature decreased and then increased, but not by how much or when.
Sometimes the shape of the graph may be all we are interested in. For this example,
however, we would like to use the graph to determine the temperature at a given time.
This will combine the specific type of information given in Table 2.1 with the ease of
seeing trends that graphs provide. Figure 2.3 displays a graph of the same time-temperature
data that includes tick marks and labeled axes. The labeling makes it clear that time is the
input and temperature is the output. The axes also show the scales used to measure each
of these. It is now possible to use the graph to estimate when the temperature decreased,
when it increased, the minimum temperature, and the maximum temperature.

Notice that the time variable is on the horizontal axis and the temperature variable
is on the vertical axis. In a function that relates time and temperature, we want time to
be the input and temperature to be the output because the temperature *depends* on the

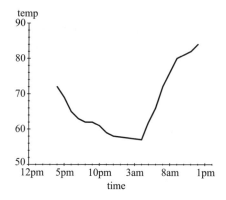

FIGURE 2.3
A relationship between time and temperature (with labels) for the data in Table 2.1.

time of day.[3] When constructing graphs, we place the input (or the independent variable) on the horizontal axis and the output (or the dependent variable) on the vertical axis. If response b depends on the value of a, we place a on the horizontal axis and b on the vertical axis. This allows us to observe the change in b as a varies. For example, the cost of a used car depends on the number of miles it has been driven; the amount of sales tax depends on the price of the purchase; how many pages of a book you read depends on how much time you spend reading. If you were to graph these relationships, the number of miles driven, the price of the purchase, and the time spent reading would be the inputs and would go on the horizontal axis.

Notice that the tick marks on the horizontal axis in Figure 2.3 are evenly spaced. For example, the distance between 5 PM and 10 PM is the same as the distance between any other five hour portion. Likewise, there is the same distance between 70° and 80° as there is between 50° and 60° or any other ten degree segment along the vertical axis. It is important to space your tick marks equally when drawing graphs. If these tick marks are not uniform, the graph will be distorted. What looks like a fast increase might, in reality, be a gradual increase. Notice in our temperature example that the same scale (i.e., distance between units) is not used on the horizontal and vertical axes. There is no reason to do so since our input and output values are measured in different units (i.e., time is in hours and temperature is in degrees). Even when the input and output values are measured in the same units, the scales do not have to be the same. Using different scales on the two axes may cause the graph to appear stretched or compressed, but there are situations where not having the same scale on both axes is advantageous.

Choosing an appropriate domain and range for graphing a function is important. If the domain or range is too small, the overall shape of the function cannot be seen. If the domain or range is too large, small changes may not be noticeable. In choosing a domain for the temperature graph, we wanted to include the entire 24-hour period. We also needed to make sure the vertical axis included the minimum temperature recorded, 56°, and the maximum temperature recorded, 84°. As you can see from Figure 2.3, we used a range that was a bit more than that. This allows a sense of "margins" for the graph. Notice that the origin[4] was not included in our temperature graph. If the origin is not part of the data, it is usually not included in the graph.

We have now seen that when data are given numerically (i.e., in a table), it is always possible to present that data graphically. We are also interested in the graphical representation of functions given symbolically. Much of the rest of this book will be concerned with combining these ideas. We will often collect or observe data of some physical phenomena. These data, when collected in the calculator as lists, can then be displayed as a graph. One question we will often ask is, "Can we find a symbolic representation that models the phenomenon?" That is, "Can we find a symbolic function that, when graphed, will have the same general shape as the data points?" If we can, we will have found a relationship that characterizes the behavior of our data.

---

[3] The temperature will also depend on your location and many other variables, but we will focus on just one input variable at a time.

[4] Remember, the origin is the point $(0, 0)$.

Words Used to Describe Graphs

Mathematicians have a vocabulary commonly used to describe graphs. Many of the words are familiar, but some may be new to you. It is important to understand the definitions of these words since discussing graphs without a common vocabulary is difficult and can lead to confusion.

Increasing and Decreasing

A function is said to be **increasing** over an interval if, for each input, a, in the interval, every input to the right of a has a greater output.[5] Symbolically, we say a function is **increasing on an interval** if for any two numbers a and b in the interval with $b > a$, then $f(b) > f(a)$. In words, this says the larger the input, the larger the output. Increasing graphs are going "uphill" as x increases toward the right. This does not have to occur over the entire graph since increasing is a property on an interval. In our temperature example (Figure 2.3), the graph is increasing on the interval from 4 AM to 12 PM.

A graph is **decreasing** over an interval if for each input, a, in that interval, every input to the right of a has a smaller output. Symbolically, we say a function is **decreasing on an interval** if for any two numbers a and b on the interval with $b > a$, then $f(b) < f(a)$. In words, this says the larger the input, the smaller the output. Decreasing graphs are going "downhill" as x increases toward the right. In our temperature example (Figure 2.3), the graph is decreasing over the interval from 4 PM to 8 PM. To determine if a graph is increasing or decreasing, we always compare outputs as we move from *left to right*.

Concavity

Concavity is another characteristic of graphs. Informally, a graph is concave up if it bends upward, resembling a bowl that can hold water. A graph is concave down if it bends downward, resembling an umbrella when it sheds water. The concavity of a graph and whether it is increasing or decreasing over an interval are independent. An increasing function can either be concave up or concave down. Similarly, a decreasing function can be either concave up or concave down. (See Figure 2.4.)

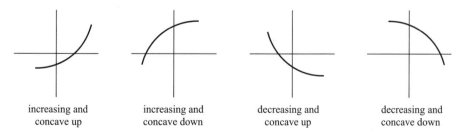

increasing and   increasing and   decreasing and   decreasing and
concave up   concave down   concave up   concave down

FIGURE 2.4
Increasing and decreasing functions with different concavity.

---

[5] The interval must be contained in the domain of the function.

The difference between increasing/decreasing and concavity is a subtle distinction. Increasing and decreasing describe how the *output* of the function is changing over an interval while the concavity of a function tells how the *rate* of increase or decrease is changing on the interval. This is connected to the definition of concavity. A graph is **concave up** on intervals where the rate of change increases. A graph is **concave down** on intervals where the rate of change decreases. For example, the graph in Figure 2.5 is always increasing. Where it is concave down (when $x < 0$), it is increasing at a slower and slower rate. So its rate of change is decreasing. Notice that the graph is less steep near the origin. Where it is concave up (when $x > 0$), it is increasing at a faster and faster rate. So its rate of change is increasing. Note that the graph gets steeper as you move to the right.

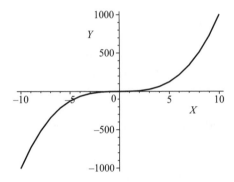

FIGURE 2.5

The graph of $y = x^3$ is concave down when $x < 0$ and concave up when $x > 0$.

Example 1. The graph of $y = x^2$ is shown in Figure 2.6. Describe where this function is decreasing, increasing, concave up, and concave down.

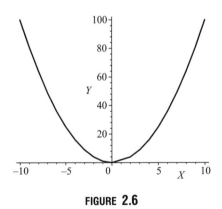

FIGURE 2.6

Solution. The function is decreasing when $x < 0$ and increasing when $x > 0$. It is always concave up and never concave down. ∎

X-intercepts and the Y-intercept

The x and y intercepts are often important points on a graph. The **y-intercept** is the y-value where the graph crosses the y-axis. This is the output when the input (or x-coordinate) of the function is zero. If b is a y-intercept, then $(0, b)$ is a point on the graph. An **x-intercept** is the x-value where the function crosses the x-axis. It is also called a *root* or *zero* of the function. The collection of values where this is true is the set of x-intercepts. They are the inputs when the output (or y-coordinate) of the function is zero. If a is an x-intercept, then $(a, 0)$ is a point on the graph. While a function can have at most one y-intercept, it can have several x-intercepts. The graph shown in Figure 2.7 has a y-intercept at $y = -6$ since it crosses the y-axis at the point $(0, -6)$. It has x-intercepts at $x = -3$ and $x = 2$ since it crosses the x-axis at the points $(-3, 0)$ and $(2, 0)$.

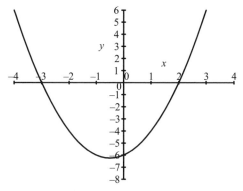

FIGURE 2.7

The graph $y = x^2 + x - 6$.

Calculator Graphics

Graphs are a good way of quickly conveying information about a function. However, graphing complicated functions by hand can be laborious. Graphing these same functions using technology can often be accomplished with a few keystrokes. The technology that makes our lives easier comes with a price (aside from what you paid for your calculator). Sometimes, we do not correctly interpret calculator produced graphs. We may just accept what we see on the screen as a good representation of our function. However, this is not always the case.

Connecting the Dots

When learning to graph functions by hand, you typically evaluate the function at various inputs, make a table containing these points, plot the points, and then "connect the dots." This is essentially how a calculator produces graphs. Graphing calculators plot points by darkening pixels on the screen. The pixels are then connected by very short lines. If your calculator evaluates the "right" points, you obtain a fairly accurate representation of the graph for your function. However, if these points do not illustrate the key properties of the function, your graph may be inaccurate.

Since a graph actually contains an infinite number of points and there are only a finite number of pixels on the calculator screen, the graph will never be completely accurate. Most of the time, however, the graph conveys enough information to correctly analyze the function. But this is not always the case. There are times when the inaccuracies of a calculator generated graph will lead you to misinterpret the function. For example, Figure 2.8 shows the screen from a graphing calculator for the graph of $f(x) = \frac{1}{x-2}$.

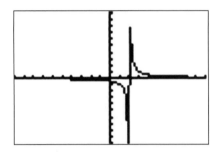

FIGURE 2.8

A TI-83 calculator graph of $f(x) = \frac{1}{x-2}$ using the default window of $-10 \le x \le 10$ and $-10 \le y \le 10$.

While this graph shows many of the important features of the function, it gives an incorrect view of what happens when $x = 2$. The graph appears to have a "spike" at $x = 2$. Actually, the function is undefined for $x = 2$ because of division by zero. There are several ways to compensate for the deficiencies in the graph. For example, once you realize that calculators will often draw nearly vertical lines where the function is undefined because of division by zero, you learn to analyze your function for these types of points and correctly sketch the graph by hand.

Suppose you wanted to use a calculator graph to determine the x-intercepts of $f(x) = x^3 + 0.2x^2$. A calculator graph of this function in the default screen is shown in Figure 2.9(a). From this graph it appears as if there is one x-intercept at approximately $(0, 0)$. We zoomed in at this point in Figure 2.9(b). It still appears as if there is one x-intercept for this function. This is actually not the case. There are two x-intercepts, one at $(-0.2, 0)$ and one at $(0, 0)$. The reason we do not see both intercepts originally is because they occur very close together at $x = -0.2$ and $x = 0$ and, between these two points, the graph goes above the x-axis by a very small amount. (See Figure 2.9(c).) The function $f(x) = x^3 + 0.2x^2$ can be analyzed symbolically by factoring. Since $f(x) = x^3 + 0.2x^2 = x^2(x + 0.2)$, the two x-intercepts are $x = 0$ and $x = -0.2$. It is important that you do not rely on your calculator for all the answers.

The Viewing Window

An important feature of a function is the shape of the graph or "the big picture." We may want to see where the function increases or decreases, if it is connected or not, or

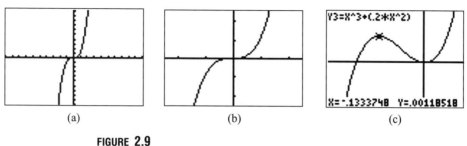

(a)　　　　　　　(b)　　　　　　　(c)

FIGURE 2.9

Three views of a calculator graph of $f(x) = x^3 + 0.2x^2$.

determine its concavity. Using a calculator graph to determine the shape of a function can be problematic since the calculator's viewing window limits what we see. In fact, if the domain of the function is "all real numbers," it is impossible to view the entire graph either with a calculator graph or a hand-drawn graph. Many graphing calculators use a default window of $-10 \le x \le 10$ and $-10 \le y \le 10$. This is not always a good choice for a viewing window. It is important to select a good viewing window for your function *before* you graph it. Do not get in the habit of always relying on the default window. To determine a proper viewing window, you need to carefully analyze your function. By knowing the general shape and important characteristics of different types of functions, you can anticipate the general shape of your function. This information helps determine an appropriate domain and range for the viewing window. It is often useful to use several different viewing windows for a single function.

Figure 2.10 shows a graph of $f(x) = x^2 + 10$ in the standard viewing window $-10 \le x \le 10$, $-10 \le y \le 10$. Notice that the screen looks blank.

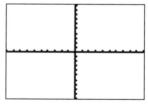

FIGURE 2.10

A graph of $f(x) = x^2 + 10$ in the standard viewing window, $-10 \le x \le 10$, $-10 \le y \le 10$.

This is because the smallest output for $f(x) = x^2 + 10$ is when $x = 0$ and $f(0) = 0^2 + 10 = 10$. Since, in the standard viewing window we only see y-values between -10 and 10, we cannot see the graph of this function. Figure 2.11 shows a revised graph of the function using a different viewing window.

Many functions have either a domain or range (or both) that includes all real numbers or is too large to use as a viewing window. There are times, however, when determining the domain and range of your function does give you the proper viewing window. This is particularly true for functions that model real world phenomena. We can often get clues from the context of the model to help determine a proper domain and range for the viewing

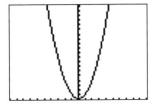

FIGURE 2.11

A revised graph of $f(x) = x^2 + 10$ using the viewing window, $-10 \leq x \leq 10$, $10 \leq y \leq 30$.

window. For example, the equation $t = 1 + \frac{s}{20} + \frac{70}{s}$ is a model used to determine the length of time a traffic light should remain yellow. In this equation, t is time (in seconds) and s is the posted speed limit (in feet per second). For this situation, you can use your driving experience to determine an appropriate viewing window. A reasonable domain for this function is 36.7 ft/sec $\approx$ 25 mph to 80.7 ft/sec $\approx$ 55 mph (the usual range of speed limits on streets where there are traffic signals). A reasonable range might be 3 to 6 seconds (a guess of how long most lights stay yellow). Figure 2.12 shows what this function looks like in the window we have chosen. While the function could be defined at speeds less than 25 mph and greater than 55 mph, it does not make sense to do so in this context.

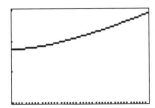

FIGURE 2.12

A graph of $t = 1 + \frac{s}{20} + \frac{70}{s}$ using a viewing window of $36.7 \leq s \leq 80.7$ and $3 \leq t \leq 6$.

Your choice of a viewing window may distort the appearance of your graph, leading to an incorrect conclusion about the function. For example, the semicircle $f(x) = \sqrt{5 - x^2}$ looks like a portion of an ellipse in the viewing window used for Figure 2.13(a), while it looks more like a semicircle in the viewing window used for Figure 2.13(b).

Some calculators have a feature that allows you to easily "square-up" a graph so that the distance between units on the x-axis is the same as the distance between units on the y-axis. (On the TI-83, this is done by using the zoom-square feature.) The common default screen of $-10 \leq x \leq 10$ and $-10 \leq y \leq 10$ does not have the same distance between units on the two axes. This default screen has the same number of *units* on the x-axis as are on the y-axis. However, since the calculator screen is wider than it is tall, the same number of units on each axis means the distance between units will be different for the two axes. This causes the graphs in the default screen to appear distorted.

Finally, there are times when you may want to adjust your viewing window so your calculator will evaluate your function at points a specified distance apart, e.g., each unit

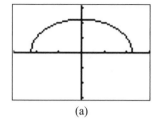   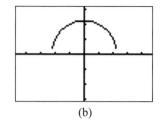

(a)   (b)

FIGURE 2.13

Graphs of $f(x) = \sqrt{5 - x^2}$ using two different viewing windows.

or each tenth of a unit. This can be thought of as the "distance" between pixels. There are different ways of creating such viewing windows. Many calculators have a zoom:integer and a zoom:decimal feature that will adjust the screen so that each pixel represents an integer or a tenth of a unit. You can also change the domain and range for your viewing window to cause the calculator to do this. In fact, you can adjust the viewing window so your function is evaluated at points any fixed distance apart. To choose a viewing window that will cause your calculator to evaluate your function at points a fixed distance apart, you need to know how wide your calculator screen is in terms of pixels and adjust the viewing window appropriately. The TI-83 viewing window has a width of 95 pixels.[6] There are 95 integers from -47 to 47 including both endpoints. Therefore, if you set your viewing window using the domain $-47 \leq x \leq 47$ and trace your graph, the function will be evaluated at every integer value. If you want it evaluated at every tenth of a unit on the TI-83, you could set your domain either from -4.7 to 4.7 or from 0 to 9.4.

Application: Analyzing a Graph of Distance Traveled by a Car

Graphs are important visual tools that help us understand the behavior of functions. That is why it is important to be able to correctly interpret graphs and draw appropriate conclusions. For example, Figure 2.14 shows the distance traveled by a car from 0 to 7 seconds.

By looking at the graph you can see that the function is increasing on the interval from 0 to 7 since the output is increasing from left to right. This means that as the input (time) increases, the output (distance) also increases. In other words, the car is moving away from the starting point. Since this graph compares distance versus time, we can make conclusions about points A, B, and C in relation to how far the car is from the starting point. When the car is at point A, it is closest to the starting point, and when it is at point C, it is the farthest from the starting point. You can also see that the graph is concave up since the shape "holds water." This means that the rate is increasing through the entire interval. In other words, the car is speeding up as it travels. So it is going the fastest at point C and the slowest at point A.

---

[6] To find the number of pixels in the screen for the calculator that you are using, look in the book that came with your calculator.

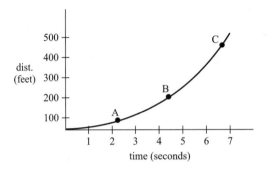

FIGURE 2.14
Graph showing distance traveled by a car from 0 to 7 seconds

Example 2.   If a car is accelerating, its velocity is increasing. If a car is decelerating, its velocity is decreasing. Is the car represented in Figure 2.14 accelerating or decelerating?

Solution.   Since the graph is concave up, the rate increases as you move from left to right on the interval. This means the velocity is increasing. Therefore, we can conclude that the car is accelerating on the interval from $t = 0$ to $t = 7$. ■

The shape of the graph tells us if the car is moving towards or away from the reference point and the concavity tells us if it is going faster or slower.

Reading Questions for Graphical Representations of Functions

1. Plot each of the given points on the grid provided. Label the point with its corresponding letter.

A. $(-1, 4)$
B. $(0, -3)$
C. $(1, -5)$
D. $(-2, 1)$
E. $(5, 2)$
F. $(-2, 0)$

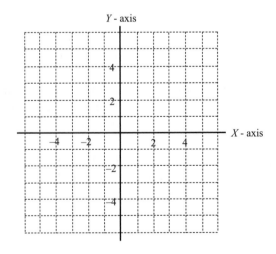

FIGURE 2.15

2. Does Figure 2.2 represent a function? Why or why not?

3. If you were to graph each pair of variables listed below as a function, which variable would go on the horizontal axis and which one would go on the vertical axis?

   (a) The amount of gasoline you put in your car *and* the cost of the gasoline.

   (b) The time it takes a car to accelerate from 0 to 60 mph *and* the horsepower of the car's engine.

   (c) Day of the year *and* amount of sunlight (i.e., time from sunrise to sunset).

   (d) A child's height *and* the child's age.

   (e) The year *and* the cost of tuition for a full-time student at your institution.

4. If someone wants to graphically de-emphasize a sudden increase or decrease, they will sometimes set the vertical scale to extend well beyond the range of the function. For example, suppose a school superintendent used a graph to represent the change in student standardized test scores from 1990 to 2002. The average standardized test score was 85 in 1990 and was 72 in 2002.

   (a) Use a vertical scale from 0 to 90 and sketch a plausible graph showing the change in average test scores.

   (b) Use a vertical scale from 70 to 85 and sketch a plausible graph showing the change in standardized test scores.

   (c) Which graph gives a more noticeable decrease? Why?

5. The graph of the following function appears to be that of a line. However, it was graphed inappropriately.

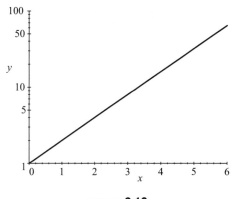

FIGURE **2.16**

   (a) What is wrong with the graph?

   (b) Sketch the graph correctly.

6. Sketch a picture of a function which is increasing and concave up for $x < 0$ and is increasing and concave down for $x > 0$.

7. Why can a function have only one y-intercept?

8. Are the "golden arches" of McDonald's concave up or concave down?

9. If the point $(4, 0)$ was on the graph of a function, would 4 be an x-intercept or the y-intercept? How do you know?

10. If you want to change the domain of your function, what setting on your graphing calculator would you change? (Be specific.)

11. Let $y = x^3 + x^2 - 4x - 4$.

    (a) Graph this function using a viewing window that allows you to easily see the three x-intercepts. What window did you use?

    (b) Find the coordinates of the three intercepts. What are they?

12. The function $f(x) = x^3 - 2x^2 + 4x - 10$ equals 5 for exactly one value of x (i.e., $x^3 - 2x^2 + 4x - 10 = 5$ for only one value of x). Find this, accurate to 3 decimal places, by graphing $y_1 = x^3 - 2x^2 + 4x - 10$ and $y_2 = 5$ and finding the intersection.

13. Use your calculator to estimate the coordinates of both intersection points of the graphs of $y = x^2$ and $y = 6 - x$.

14. Use your calculator to graph each of the following. Sketch the graph and describe it using the terms *y-intercept, x-intercept, increasing, decreasing, concave up, concave down*.

    (a) $y = x^2 - 1$. Use the viewing window $-4 \le x \le 4$, $-2 \le y \le 16$.

    (b) $y = x^3 + \frac{1}{2}x^2 - 2x$. Use the viewing window $-4 \le x \le 4$, $-6 \le y \le 6$.

    (c) $y = 3^x - 1$. Use the viewing window $-2 \le x \le 2$, $-1 \le y \le 9$.

    (d) $y = \frac{1}{(x-1)}$. Use the viewing window $-1 \le x \le 3$, $-8 \le y \le 8$.

15. Given the function $f(x) = \frac{1}{(x-1)}$, answer the following questions.

    (a) Algebraically, what is happening when $x = 1$?

    (b) What does a calculator produced graph of $y = \frac{1}{(x-1)}$ look like at $x = 1$?

16. Why is it important to choose an appropriate viewing window when using a calculator to graph a function?

17. Graph $f(x) = 4x^3 - 0.4x^2$ in the default screen $-10 \le x \le 10$, $-10 \le y \le 10$.

    (a) How many x-intercepts do you see?

    (b) There are actually two x-intercepts. Factor $4x^3 - 0.4x^2 = 0$ to find the x-intercepts.

    (c) Explain why you are unable to see both x-intercepts in your calculator graph.

18. Let $f(x) = 0.005x^3 - 0.12x^2 + 8$.

    (a) Graph f with your viewing window set at $-7 \le x \le 10$ and $-10 \le y \le 10$. Sketch the graph as it appears in this viewing window.

    (b) Graph f with your viewing window set at $-10 \le x \le 25$ and $-10 \le y \le 10$. Sketch the graph as it appears in this viewing window.

(c) Which of your two sketches from the above questions best describes the shape of the function? Why?

19. A standard calculator screen of $-10 \leq x \leq 10$ and $-10 \leq y \leq 10$ is not an appropriate window to view the general shape of the graphs of the functions given below. Analyze each of the following functions to determine an appropriate viewing window that shows the graph's general shape and make a sketch of your calculator graph. Be sure to indicate the dimensions of the viewing window you used.

(a) $f(x) = x + 20$

(b) $h(x) = (x - 10)^2$

20. The following graph represents the distance traveled by a car. Use the information represented by the graph to answer the following questions.

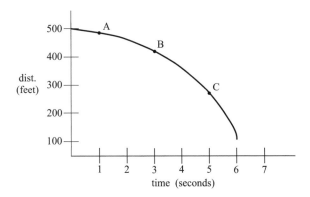

FIGURE 2.17

(a) The graph is decreasing. What does that tell us about the car?

(b) The graph is concave down. What does that tell us about the car?

(c) At which point is the car farthest away from the point of reference?

(d) At which point is the car going the fastest?

Graphical Representations of Functions: Activities and Class Exercises

1. **Pneumonia Graph.** Use Figure 2.18 and the excerpt from the article on *U.S. Deaths from Pneumonia*[7] to answer the following questions.

   "Pneumonia probably affected prehistoric humans and is one of the oldest diagnosed diseases, having been described by the Hippocratic physicians of

---

[7] "U.S. Deaths from Pneumonia," *Scientific American*, Feb. 1997, p. 29.

ancient Greece. In 1900 it was the second deadliest killer in the US after tuberculosis. The extraordinary high rate in 1918 resulted from the great influenza epidemic that year, which killed more than 540,000 Americans. Since then, pneumonia mortality rates have decreased markedly because of better hygiene and increasingly effective methods of treatment: first, antip-neumococcal serum, then sulfa drugs, and finally, in the 1940s, penicillin. The increase in deaths during the past 15 years stems primarily from the growing number of old people."

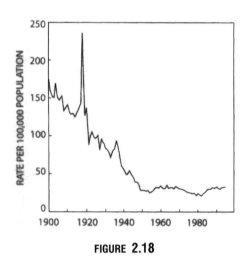

FIGURE 2.18

(a) Why is there a spike in the graph close to 1920?

(b) After 1920, the graph is basically decreasing, but starts to slowly increase after 1980. What is causing this increase?

(c) Sketch what you expect the graph to look like from 1990 to 2010. What assumptions are you making?

(d) Why is the output in "rate per 100,000 population" rather than in total deaths from pneumonia?

(e) According to the Census Bureau, the population in 1900 was approximately 76,212,168. Approximately how many people died of pneumonia that year?

(f) According to the Census Bureau, the population in 1980 was approximately 226,542,203. Approximately how many people died of pneumonia that year?

2. **Examples of Poor Graphs.** Look at the three graphs (Figure 2.19) titled "How Vote Getters Voted," "Advil," and "OttawaRide Ridership."

   For each graph, answer the following questions.

   (a) What impression does the graph convey?

   (b) What is wrong with the graph?

   (c) If the graph were drawn correctly, what impression would it most likely convey?

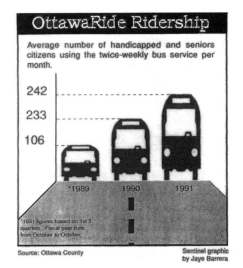

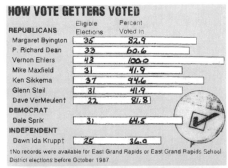

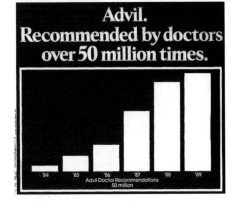

FIGURE 2.19

3. **Newspaper Search.** Using a newspaper or magazine, do the following:

   (a) How many times did the newspaper or magazine display information graphically?

   (b) For each graph, indicate what it represented. Give the domain and range.

   (c) Does your newspaper or magazine appear to use graphs appropriately? Why or why not?

4. **Overweight Americans.** Look at the graph for *Overweight Americans* taken from the *New York Times* (Figure 2.20). This graph compares percentages of overweight and obese Americans. The information is organized by gender as well as time period. "Overweight" is classified as 25 to 30 pounds over the recommended weight, while "obese" is classified as more than 30 pounds over the recommended weight.

   (a) Compare the size of the bar for obese women from 1991 to 1994 to the size of the bar for obese women from 1976 to 1980. What conclusion can you draw?

   (b) If you randomly choose 100 men in 1992, how many would be obese, on average?

   (c) The population of the United States in 1980 was approximately 226,542,203. Estimate the total number of obese people that year, including both men and women.

   (d) We are interested in comparing the rate of increase of obese men. We notice that there is a small change in the height comparing the first bar (71–74) to the second bar (76–80) while there is a large change in the height from the second

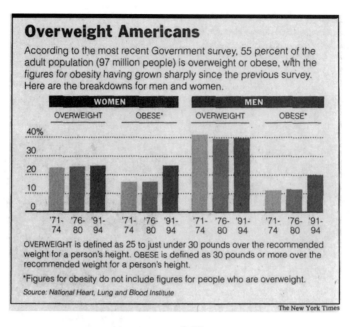

Overweight Americans

According to the most recent Government survey, 55 percent of the adult population (97 million people) is overweight or obese, with the figures for obesity having grown sharply since the previous survey. Here are the breakdowns for men and women.

OVERWEIGHT is defined as 25 to just under 30 pounds over the recommended weight for a person's height. OBESE is defined as 30 pounds or more over the recommended weight for a person's height.

*Figures for obesity do not include figures for people who are overweight.

Source: National Heart, Lung and Blood Institute

The New York Times

FIGURE **2.20**

bar to the third bar (91–94). From this we conclude that the rate has dramatically increased. Explain what is wrong with this reasoning.

5. **Dance Injuries.** Look at Figure 2.21 and use it to answer the following questions.

   (a) For the body parts given in the graph, which one appears to be at least risk for dancers?

   (b) Estimate the total number of injuries reported in this study.

   (c) If a dancer has an injury to a lower extremity, what is the likelihood that it is a knee injury? Your answer should be expressed as a percent.

   (d) The number of injuries reported is greater than the number of dancers in the study. Provide an explanation for this observation.

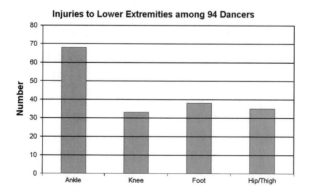

FIGURE **2.21**

6. **Attendance at Museum.** The graph in Figure 2.22 gives attendance at the Nova Scotia Museum in 1997. Use it to answer the following questions.

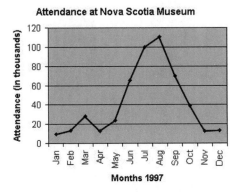

FIGURE 2.22

(a) There is a spike in the graph for August. What does this mean in terms of attendance?

(b) Approximately how many people attended the museum in September of 1997?

(c) The museum needs to hire one employee per one thousand visitors. How many employees, on average, will be needed for the summer months (June, July, and August)?

(d) Attendance figures for 1997 indicate increases in attendance in March and during the summer months. Would you expect the same pattern to hold true for 1998? Why or why not?

7. **Prison Population.** The graph, in Figure 2.23, titled *Distribution of Offenses: State and Federal Prisons*[8] lists the percentage of the prison population incarcerated for drug offenses, property offenses (including burglary, larceny, and fraud), violent offenses, public order offenses, and other. The data compares 1985 to 1995 as well as state prisons to federal prisons.

(a) Why is each bar in the graph the same height?

(b) What additional information is needed to determine the actual number of people incarcerated in federal prisons because of property offenses in 1985?

(c) The percentage of public order offenses in federal prisons in 1995 (18.3%) was approximately twice as much as public order offenses in federal prisons in 1985 (9.1%). Write a similar statement comparing drug offenses in federal prisons.

(d) Considering the change in drug offenses in federal prisons from 1985 to 1995 (see your answer to part (c)), explain why it is impossible for a trend of similar magnitude to hold true from 1995 to 2005.

---

[8] Haney, Craig and Philip Zimbardo, "Distribution of Offenses: State and Federal Prisons," *American Psychologist*, Vol. 53. No. 7, p. 715.

Distribution of Offenses: State and Federal Prisons, 1985 and 1995

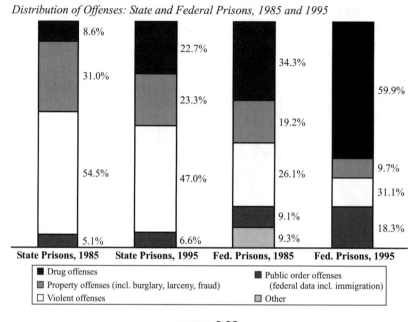

■ Drug offenses	■ Public order offenses
▦ Property offenses (incl. burglary, larceny, fraud)	(federal data incl. immigration)
☐ Violent offenses	☐ Other

FIGURE **2.23**

8. **Species Introduced into the Great Lakes.** The graph[9] in Figure 2.24 depicts the type of non-native species introduced into the Great Lakes, organized by 30 year periods.

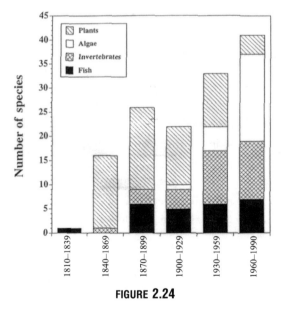

FIGURE **2.24**

---

[9] Mills, Edward, et al., "Exotic species and the integrity of the Great Lakes," *BioScience*, Vol. 44, No. 10, p. 668.

(a) During which time period was the total number of non-native species introduced to the Great Lakes the greatest?

(b) The annual rate of introduction of non-native species from 1930 to 1959 is about how many species per year?

(c) How many more non-native algae species were introduced during the period from 1960 to 1990 as compared to the period from 1900 to 1929?

(d) Which type of these three groups (plants, invertebrates or fish) have experienced the greatest increase in the **rate** of introductions since 1870?

9. **Sleep Cycles.** The graph in Figure 2.25 compares sleep cycles for children, young adults, and the elderly. Stage 4 is the deepest sleep stage and REM (Rapid Eye Movement) is the shallowest sleep stage.

(a) How many times per night do young adults typically awaken?

(b) The dark areas in the chart represent what sleep stage? Explain how you came to your conclusion.

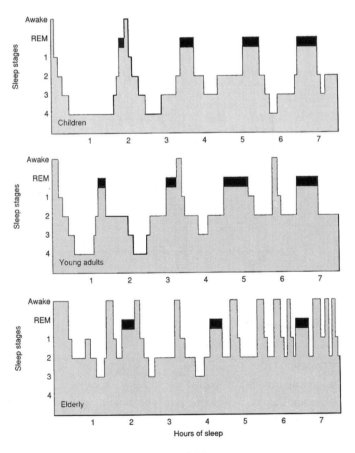

FIGURE **2.25**

(c) Estimate, to the nearest 10 minutes, the time the typical young adult was awake during the $7\frac{1}{2}$ hours of sleep depicted in the chart.

(d) Nightmares are most likely to occur during REM sleep. Using the information in the chart, are children more, less, or just as likely to have nightmares compared to young adults? Justify your answer, being as specific as possible.

10. **Gasoline Prices.** Below is a table and graph (Figure 2.26) that gives the average gasoline prices from 1990 to 2002.[10]

(a) Where is the graph increasing? Where is it decreasing?

(b) Where is it concave up? Where is it concave down?

(c) What does it mean, in terms of gas prices, if the graph is concave up?

(d) What do you find most surprising about this graph?

Year	1990	1991	1992	1993	1994	1995	1996
Price	1.16	1.14	1.13	1.11	1.11	1.15	1.23

Year	1997	1998	1999	2000	2001	2002
Price	1.23	1.06	1.17	1.51	1.46	1.23

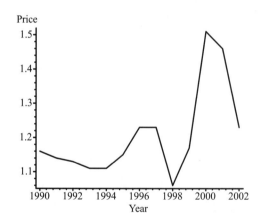

FIGURE 2.26

(e) Using this information, do you think you can accurately predict the price of gasoline in 2005? Why or why not?

11. **Weather Trends.** The function $T(m) = 0.0235m^4 - 0.678m^3 + 5.43m^2 - 8.23m - 11.1$ can be used to model the average monthly temperature in **Celsius** for Fargo, North Dakota.[11] The input is the month where 1=January, 2=February,...,12=December.

---

[10] "Public Data Query," 10 June 2002, <http://data.bls.gov/servlet/SurveyOutputServlet>.
[11] "Averages and Records," 12 March 2004, <http://www.weather.com/weather/climatology/USND0115>.

(a) What is an appropriate viewing window for this function?

(b) Graph this on your calculator and draw a sketch.

(c) Find the coordinates of the maximum. What does this tell you about the temperature in Fargo, North Dakota?

(d) Find the x-intercepts. What do these points represent in terms of temperature?

(e) Seventy degrees Farenheit is approximately equal to 21.1 degrees Celsius. Find the coordinates of the point(s) where the temperature is 70 degrees Farenheit. What does this tell you about the temperature in Fargo, North Dakota?

3

Applications of Graphs

In the last section, we looked at graphs and the words used to describe them. However, graphs are typically used to describe some physical phenomenon. For example, a graph may represent the average daily temperature, the daily cost of a gallon of gasoline, or the progression of a child's fever. When a graph is given in a physical context, its shape and properties also have physical meaning. In this section, we will look at graphs and what they represent in physical contexts. We will also look at how to shift graphs both vertically and horizontally.

Using Graphs to Represent Physical Situations

Graphical representations of functions are only useful if you are able to correctly interpret the information they represent. Therefore, an important skill is being able to convert a verbal description of a physical situation to a graph and vice versa. Often, we are primarily interested in the general shape of the graph. In this case, the only labels included in the graph may be descriptions of the inputs and outputs. As an example, suppose a person was walking away from you at a constant speed and you wanted to graphically represent this situation. The two variables are time and distance. Since distance *depends* on time, a graph of this situation would have time on the x-axis and distance on the y-axis. The graph representing the function would be an increasing line (See Figure 3.1(a).) The graph is increasing because the person is walking away from you. The graph is a line because the person is walking at a constant speed. If the graph were a horizontal line, the output would not be changing. This would mean that the person would not be moving. (See Figure 3.1(b)). If the graph was a decreasing line, the person would be moving closer to you. (See Figure 3.1(c)).

What if the person was not walking at a *constant rate*? In modeling physical situations, you need to consider not only whether the function is increasing or decreasing, but the rate of increase or decrease as well. This means concavity is important. Figure 3.2 contains

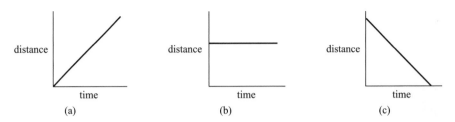

FIGURE 3.1
(a) represents a person walking away from you at a constant rate. (b) represents a person standing still. Figure (c) represents a person walking towards you at a constant rate.

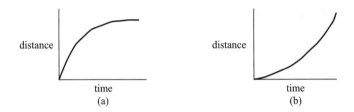

FIGURE 3.2
Two graphs showing a person's position as he or she walks down a hallway. In (a), the person is walking more slowly as they move away. In (b), the person is walking more quickly as they move away.

two graphs representing the distance a person is away from you as he or she walks down a hallway.

In Figure 3.2(a), the person starts out quickly walking away from you and then slows down. This gives a graph that is always increasing, but is concave down since the person's rate of walking is decreasing. In Figure 3.2(b), the person starts out slowly walking away from you and then speeds up. This gives a graph that is also increasing, but is concave up since the person's rate of walking is increasing. Remember that concavity shows how the person's *rate* of walking is changing. A graph that is concave up will have a rate of change that is increasing while a graph that is concave down will have a rate of change that is decreasing. Notice that both graphs are increasing over the entire interval, showing that the person is always walking away from you.

Example 1. The graph shown in Figure 3.3 represents a function where time is the input and the distance a person is away from you as he or she walks down a hall is the output. Describe how that person walked.

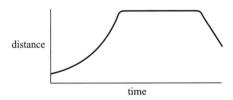

FIGURE 3.3

Solution.   Since the graph is increasing and concave up at the beginning, the person is walking away from you and the rate of walking is increasing. Thus, the person starts out walking slowly away and then speeds up. Where the graph is horizontal, the person has stopped since the distance remains constant. The stop occurs over a measurable time frame—exactly that time for which the graph is flat. Where the graph is decreasing, the person is walking back towards you at a constant rate. Note that this piece of the graph is a straight line. Since the graph stops at a higher point than where it began, the person stopped before returning to the starting point. ■

Example 2.   Suppose you threw a baseball straight up in the air so that its maximum height was 30 feet and caught it as it returned to the ground. Describe how the height of the ball changed over time and sketch a graph of this situation where time is the input and the height of the ball is the output.

Solution.   You would have to throw a ball fairly hard to get it to a height of 30 feet. So initially the ball would travel quite rapidly. However, gravity would cause it to slow down until, at 30 feet, it would actually stop. This is the vertex (or maximum point) of the graph. It will then start to fall back towards the ground. As it does, gravity will cause it to speed up until you catch it. A graph of this situation, with time as the input and height above the ground as the output, is shown in Figure 3.4.

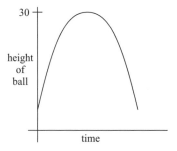

FIGURE 3.4

Notice that this is **not** a picture of the flight of the ball—that would be a vertical line. It is a graph of the height of the ball versus time. The shape of the graph corresponds to our verbal description. When the graph is steep, it indicates that the ball is traveling rapidly. This occurs soon after the ball is thrown and just before it is caught. In the middle, the graph is not as steep. This indicates that the ball was traveling more slowly. ■

The last two examples involved motion. But graphs can describe other situations as well. Suppose you were filling a common cylinder-shaped drinking glass with water from a faucet and measuring the height of the water in the glass. The water level in the glass will rise at a constant rate. The graph of this event, with time as the input and height of water as the output, would look like an increasing line since the water level is rising and rate of increase is constant. (See Figure 3.5.)

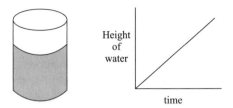

FIGURE 3.5
If water is added to a cylinder at a constant rate, the graph describing the water will be linear.

Example 3.   Suppose you are filling a glass in the shape of a cone with water from a faucet. (See Figure 3.6.) Sketch a graph of the function whose input is time and whose output is the level of water.

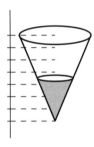

FIGURE 3.6

Solution.   The graph will be similar to Figure 3.7.

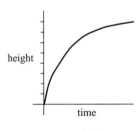

FIGURE 3.7

Since the radius of the cone near the bottom is very small, the water level will rise very quickly at first. However, as we continue to add water, the rate at which the level is rising slows down since it takes more water for the level to go up near the top of the cone. The graph is increasing since the height of the water increases as the glass is filled. The graph starts out quite steep (since we start out with the water level rising quickly) and gets less steep as we approach the top of the glass. This will give us a graph that is *concave down* for the entire interval. Our graph will start at the origin since the glass is empty when we begin. The domain of our function ends when the glass is full. ∎

Shifting Functions

Adding a Constant to the Output

Often, we have the general shape of a function, but need to "shift" it. We will concentrate on the graphical representations of shifted functions. However, shifting works regardless of how the function is defined (i.e., symbolically, graphically, numerically or even verbally). As we study particular functions later in the course, we will revisit the concepts introduced here. As an example, consider $f(x) = x^2$ and $g(x) = x^2 + 5$. Both functions have the same shape but the second one is shifted 5 units higher. When 5 is added to the output, the y-values increase by 5, thereby raising the graph. For instance, if $x = 1$, then $f(1) = 1^2 = 1$ and $g(1) = 1^2 + 5 = 1 + 5 = 6$, which is 5 units higher. (See Figure 3.8.)

Contrast the function $h(x) = 5x^2$ with $f(x) = x^2$. In this case the output is increased by a multiple of 5. In Figure 3.9(b), you can see that $h(x) = 5x^2$ is vertically stretched

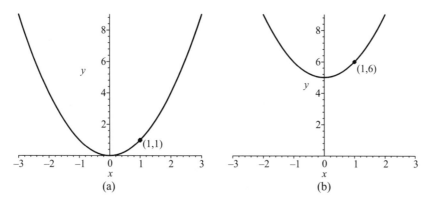

FIGURE 3.8

Graph (a) is $y = x^2$ and Graph (b) is $y = x^2 + 5$. Notice that the second graph has the same shape, but is shifted 5 units higher.

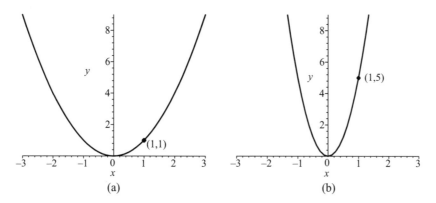

FIGURE 3.9

Graph (a) is $y = x^2$ and Graph (b) is $y = 5x^2$. Notice that the second graph is vertically stretched by a factor of 5.

by a factor of 5. Again, if $x = 1$, then $f(1) = 1^2 = 1$ and $h(1) = 5 \cdot 1^2 = 5 \cdot 1 = 5$, which is increased by a factor of 5. Both $g(x) = x^2 + 5$ and $h(x) = 5x^2$ involved changes to the output and therefore change the graph vertically, but in different ways.

The function $f(x) = (x + 5)^2$ adds 5 to the input. In this case, the graph is shifted horizontally. (See Figure 3.10(b)). Notice that adding 5 to the input shifts the graph to the *left*.

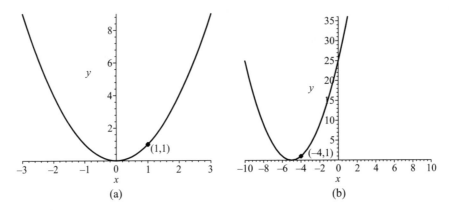

FIGURE 3.10

Graph (a) is $y = x^2$ and Graph (b) is $y = (x + 5)^2$. Notice that the second graph is shifted horizontally 5 units to the *left*.

Finally, $f(x) = (5x)^2$ results in a horizontal compression as shown in Figure 3.11(b). Let's now consider a more complicated function. Let $y = w(x)$ be the function shown in Figure 3.12. This graph could have been generated by using a motion detector and having a person walk as indicated in the boxed instructions.

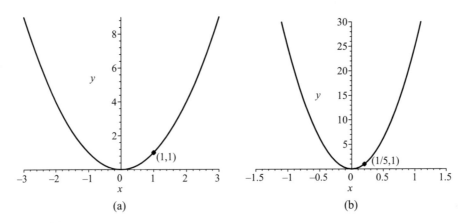

FIGURE 3.11

Graph (a) is $y = x^2$ and Graph (b) is $y = (5x)^2$. Notice that the second graph is compressed horizontally by a factor of 5.

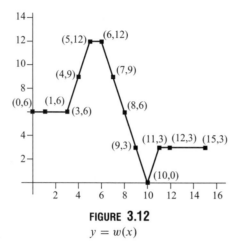

FIGURE 3.12
$y = w(x)$

Directions for using a motion detector to produce $y = w(x)$

- Start 6 feet away from the motion detector.
- Stand still for 3 seconds.
- Step 6 feet away from the starting point, stepping 3 feet each second.
- Stand still for 1 second.
- Walk toward the motion detector for 4 seconds, moving 3 feet each second.
- Step 3 feet away from the motion detector (stepping 3 feet per second).
- Stand still for the remaining 4 seconds.

The domain of $y = w(x)$ is $0 \le x \le 15$ and the range of $y = w(x)$ is $0 \le y \le 12$. Several points are labelled on the graph. We are now going to define a new function: $g(x) = w(x) + 2$. Let us think about what this notation means. For any x in the domain of $y = g(x)$, we find the associated y-value by first finding the y-value of $y = w(x)$ and then adding 2. So the domain values of $y = g(x)$ are the same as the domain values

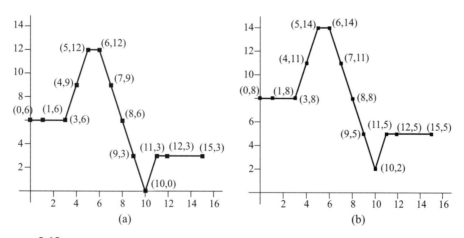

FIGURE 3.13
Graph (a) is $y = w(x)$ and Graph (b) is $y = w(x) + 2$. The second graph has been shifted up 2 units.

of $y = w(x)$. Figure 3.13(b) is the graph of the new function $y = g(x)$. Notice that the points labelled in the graph of $y = g(x)$ are two units higher than the points labelled on the graph for $y = w(x)$.

The shape of the graph for $y = g(x)$ is the same as the shape of the graph for $y = w(x)$. It has just been shifted 2 units higher. To create this graph by walking, you just start 8 feet (2 feet more than $y = w(x)$) away from the motion detector and follow the same directions.

Directions for using a motion detector to produce $y = w(x)$	Directions for using a motion detector to produce $y = w(x) + 2$
• Start 6 feet away from the motion detector.	• Start **8** feet away from the motion detector.
• Stand still for 3 seconds.	• Stand still for 3 seconds.
• Step 6 feet away from the starting point, stepping 3 feet each second.	• Step 6 feet away from the starting point, stepping 3 feet each second.
• Stand still for 1 second.	• Stand still for 1 second.
• Walk toward the motion detector for 4 seconds, moving 3 feet each second.	• Walk toward the motion detector for 4 seconds, moving 3 feet each second.
• Step 3 feet away from the motion detector (stepping 3 feet per second).	• Step 3 feet away from the motion detector (stepping 3 feet per second).
• Stand still for the remaining 4 seconds.	• Stand still for the remaining 4 seconds.

The function $h(x) = w(x) - 2$ is similar except it is shifted *down* 2 units. The graph for $y = h(x)$ is shown in Figure 3.14(b).

In general, given a function $y = f(x)$, we can define a new function as $y = f(x) \pm c$.[1] (This notation means we are either adding a constant number "c" or subtracting a constant number "c" to the output $f(x)$ where "$\pm$" is read as "plus or minus.") The graph of the new function $y = f(x) \pm c$ is just the graph of the old function $y = f(x)$ shifted up if c is added or shifted down if c is subtracted.

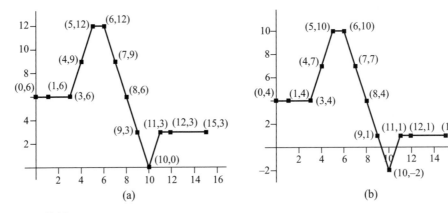

(a)   (b)

FIGURE 3.14
Graph (a) is $y = w(x)$ and Graph (b) is $y = w(x) - 2$. The second graph has been shifted down 2 units.

---

[1] We are assuming c is positive.

Multiplying the Output by a Constant

Suppose we want to multiply a function by a constant. Consider $k(x) = 2w(x)$. This means we repeat the steps in making the graph of $y = w(x)$, but walk twice as fast.

Directions for using a motion detector to produce $y = w(x)$	Directions for using a motion detector to produce $y = 2w(x)$
• Start 6 feet away from the motion detector.	• Start **12** feet away from the motion detector.
• Stand still for 3 seconds.	• Stand still for 3 seconds.
• Step 6 feet away from the starting point, stepping 3 feet each second.	• Step **12** feet away from the starting point, stepping **6** feet each second.
• Stand still for 1 second.	• Stand still for 1 second.
• Walk toward the motion detector for 4 seconds, moving 3 feet each second.	• Walk toward the motion detector for 4 seconds, moving **6** feet each second.
• Step 3 feet away from the motion detector (stepping 3 feet per second).	• Step **6** feet away from the motion detector (stepping **6** feet per second).
• Stand still for the remaining 4 seconds.	• Stand still for the remaining 4 seconds.

For an x in the domain of $y = k(x)$, we find the associated range value by first finding the range value of $y = w(x)$, then multiplying that value by 2. This implies that the domain values of of $y = k(x)$ are the same as the domain values of $y = w(x)$ and the associated range values of $y = k(x)$ are 2 times the range values of $y = w(x)$. A graph of $y = k(x)$ is given in Figure 3.15(b). Notice that each point is twice as high as the corresponding points for $y = w(x)$.

We can see that the graph of $y = k(x)$ is a vertically "stretched" version of the original graph where any points on the x-axis, such as the point $(10, 0)$, stay in the same place. If we imagine the original graph of $y = w(x)$ on a sheet of rubber, the effect is as if we

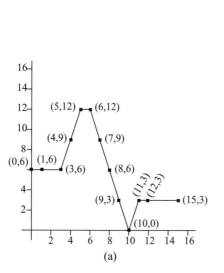

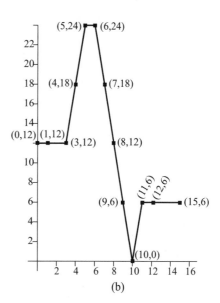

(a)

(b)

FIGURE 3.15
Graph (a) is $y = w(x)$ and Graph (b) is $y = 2w(x)$. The second graph has been vertically stretched by a factor of 2.

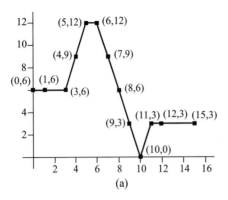 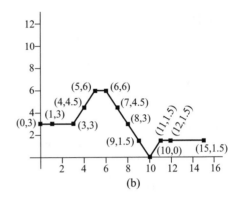

(a)   (b)

FIGURE 3.16

Graph (a) is $y = w(x)$ and Graph (b) is $y = \frac{1}{2}w(x)$. The second graph has been vertically compressed by a factor of 2.

glued the x-axis in place, then pulled the sheet upward by moving each point twice as far away from the x-axis. The function $j(x) = \frac{1}{2}w(x)$ is similar except it is *compressed* by a factor of 2. This would be similar to walking half as fast (or twice as slow). (See Figure 3.16.)

In general, defining a new function as $y = cf(x)$ will stretch or compress a function away (if $c > 1$) or toward (if $0 < c < 1$) the x-axis.

Adding a Constant to the Input

So far, we have been changing the output, y, by adding a constant or multiplying. But what if we want to change the input, x, instead? Let $p(x) = w(x + 2)$.

This means that the time (input) starts two units *earlier*. Consider what is happening. For a value x in the domain of $y = p(x)$, first "compute" $x + 2$. Then, using $x + 2$ as **a domain value for $y = w(x)$**, find $w(x + 2)$. Remember that the domain of $y = w(x)$ is 0 to 15, so now we must have $0 \le x + 2 \le 15$. This means $-2 \le x \le 13$. So the domain of $y = p(x)$ will be $-2 \le x \le 13$ because this causes $x + 2$ to be in the domain of $y = w(x)$. Let us look at some particular points on the graph of $y = p(x)$. Start with $x = -2$.

Directions for using a motion detector to produce $y = w(x)$	Directions for using a motion detector to produce $y = w(x + 2)$
	• Start **2 seconds earlier** than the person making the graph for $w(x)$
• Start 6 feet away from the motion detector.	• Start 6 feet away from the motion detector.
• Stand still for 3 seconds.	• Stand still for 3 seconds.
• Step 6 feet away from the starting point, stepping 3 feet each second.	• Step 6 feet away from the starting point, stepping 3 feet each second.
• Stand still for 1 second.	• Stand still for 1 second.
• Walk toward the motion detector for 4 seconds, moving 3 feet each second.	• Walk toward the motion detector for 4 seconds, moving 3 feet each second.
• Step 3 feet away from the motion detector (stepping 3 feet per second).	• Step 3 feet away from the motion detector (stepping 3 feet per second).
• Stand still for the remaining 4 seconds.	• Stand still for the remaining 4 seconds.

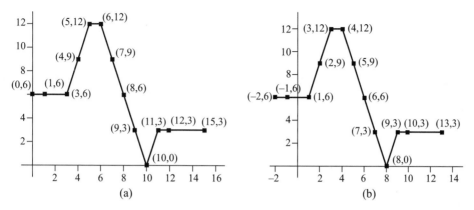

FIGURE 3.17

Graph (a) is $y = w(x)$ and Graph (b) is $y = w(x + 2)$. The second graph has been shifted 2 units to the **left**.

Then $p(-2) = w(-2 + 2) = w(0) = 6$. So the value of $y = p(x)$ at its "left end" domain value of -2 is the value of $y = w(x)$ at its "left end" domain value (which is 0). The point $(0, 6)$ on the graph of $y = w(x)$ is shifted two units to the left to get the value $(-2, 6)$ on the graph of $y = p(x)$. In fact, the graph of $p(x) = w(x + 2)$ is just the graph of $y = w(x)$ shifted 2 units to the left as shown in Figure 3.17. To make this graph with a motion detector, follow the same steps as for $y = w(x)$ but start walking 2 seconds earlier.

The graph of $t(x) = w(x - 2)$ is shown in Figure 3.18(b). Notice that it is the graph of $y = w(x)$ shifted 2 units to the right. To make this graph with a motion detector, follow the same steps as for $y = w(x)$ but walk 2 seconds later.

In summary, we can see that defining a new function as $f(x + c)$ or $f(x - c)$ will shift the graph to the left (if it is $+c$) or right (if it is $-c$) by c units.[2]

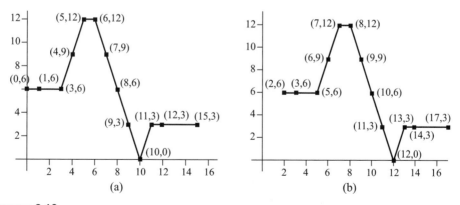

FIGURE 3.18

Graph (a) is $y = w(x)$ and Graph (b) is $y = w(x - 2)$. The second graph has been shifted 2 units to the **right**.

---

[2] We are assuming c is positive.

Multiplying the Input by a Constant

Let us next consider defining a function as $r(x) = w(2x)$. In this case the *time*, rather than the speed, is going twice as fast.

Directions for using a motion detector to produce $y = w(x)$	Directions for using a motion detector to produce $y = w(2x)$
• Start 6 feet away from the motion detector.	• Start 6 feet away from the motion detector.
• Stand still for 3 seconds.	• Stand still for **1.5** seconds.
• Step 6 feet away from the starting point, stepping 3 feet each second.	• Step 6 feet away from the starting point, stepping 3 feet each **0.5** second.
• Stand still for 1 second.	• Stand still for **0.5** second.
• Walk toward the motion detector for 4 seconds, moving 3 feet each second.	• Walk toward the motion detector for **2** seconds, moving 3 feet each **0.5** second
• Step 3 feet away from the motion detector (stepping 3 feet per second).	• Step 3 feet away from the motion detector (stepping 3 feet per **0.5** second).
• Stand still for the remaining 4 seconds.	• Stand still for the remaining **2** seconds.

For any value x **in the domain of** $y = r(x)$ we first find $2x$, then find $w(2x)$. Since the domain of $y = w(x)$ is 0 to 15, we use x values so that $2x$ is in the domain of w, i.e., $0 \le 2x \le 15$. This means $0 \le x \le 7.5$. The graph of $y = r(x)$ is shown in Figure 3.19(b). Its domain is 0 to 7.5 and its range is 0 to 12 (the same as $w(x)$).

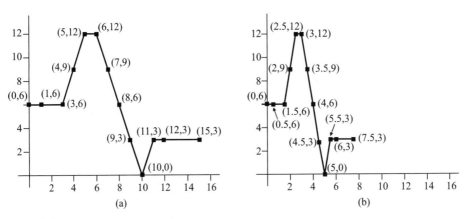

FIGURE 3.19

Graph (a) is $y = w(x)$ and Graph (b) is $y = w(2x)$. The second graph has been horizontally compressed by a factor of 2.

The graph of $y = r(x)$ is a compressed version of the graph of $y = w(x)$. When $r(x) = w(2x)$, the y-axis is fixed and there is a horizontal compression toward the y-axis. The function $s(x) = w\left(\frac{1}{2}x\right)$ is similar except time is going half as fast. In this case, the graph is horizontally stretched by a factor of 2. (See Figure 3.20(b)).

In general, the graph of $y = w(cx)$, will be horizontally compressed if $c > 1$ and horizontally stretched if $0 < c < 1$.

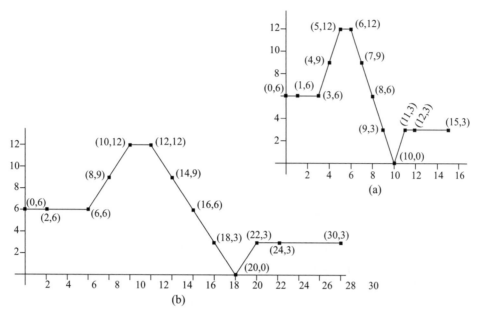

FIGURE 3.20

Graph (a) is $y = w(x)$ and Graph (b) is $y = s(x) = w(\frac{1}{2}x)$. The second graph has been horizontally stretched by a factor of 2.

Summary of Shifts and Stretches

In summary, **output changes** for $y = f(x)$ include:

- $y = f(x) + c$ is the graph of $y = f(x)$ shifted up c units (assuming $c > 0$).
- $y = f(x) - c$ is the graph of $y = f(x)$ shifted down c units (assuming $c > 0$).
- $y = cf(x)$ is the graph of $y = f(x)$ vertically stretched by a factor of c if $c > 1$.
- $y = cf(x)$ is the graph of $y = f(x)$ vertically compressed by a factor of $1/c$ if $0 < c < 1$.

Input changes include:

- $y = f(x + c)$ is the graph of $y = f(x)$ shifted to the **left** c units (assuming $c > 0$).
- $y = f(x - c)$ is the graph of $y = f(x)$ shifted to the **right** c units (assuming $c > 0$).
- $y = f(cx)$ is the graph of $y = f(x)$ horizontally compressed by a factor of c if $c > 1$.
- $y = f(cx)$ is the graph of $y = f(x)$ horizontally stretched by a factor of $1/c$ if $0 < c < 1$.

Application: Changing the Thermostat

We shift a function when we want to preserve a function's shape but change where it begins and ends. As an example, consider an automatic thermostat that is used to turn

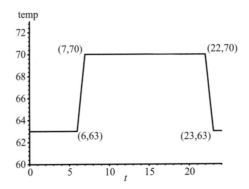

FIGURE 3.21
This graph represents the temperature in a house during a 24 hour period. Time 0 is 12 AM.

the furnace on and off in the winter. Having the furnace heat y more degrees or turning on x hours earlier is equivalent to shifting a function.

Many homes have automatic thermostats for the heating and cooling systems. These thermostats are programmed to keep the house at a preferred temperature during specified times of the day. The following graph represents this situation. The number of hours since midnight is the input and the temperature of a house, in degrees Fahrenheit, is the output. Some key points on the graph are included. Time 0 represents 12 AM. Looking at Figure 3.21, we see that for 6 hours after 12 AM, the graph is horizontal at a height of 63. We can deduce that the thermostat is set at 63 degrees at night. At 6 AM, the thermostat turns on the furnace and the temperature increases to 70 over the next hour. At 7 AM, the graph is horizontal with a height of 70. It stays at 70 degrees until 22 hours or 10 PM. We can deduce that the thermostat is set at 70 degrees during the day from 6 AM to 10 PM. At 10 PM, the furnace turns off and the temperature decreases to 63 degrees at 11 PM.

Example 4.   Suppose you think 63 degrees at night and 70 degrees during the day is too chilly and you decide to raise the temperature in the house by 3 degrees. What would be your nighttime temperature (at 12 AM) and daytime temperature (at 7 AM)? What would the new graph look like?

Solution.   In order to visually see this based on our original graph, we need to refer to what we learned in the shifting function section. Since we are adjusting temperature we are adding a constant to the output. This will raise the graph 3 units higher as shown in Figure 3.22. Our nighttime temperature is now 66 degrees and our daytime temperature is now 73 degrees. ■

In Example 4, we looked at changes in the graph. If we are interested in the symbolic form of the function, we need to first realize that we are changing the *output*. So, if $y = h(t)$ is the original function, either $y = h(t) + c$ or $y = ch(t)$ will be the new function. Since we are adding 3 degrees to the temperature (our output), this is a shift rather than a compression. Therefore, our new function will be of the form $y = h(t) + c$. Finally, we are adding *three* degrees, so our new function is $y = h(t) + 3$.

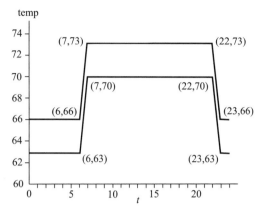

FIGURE **3.22**

Reading Questions for Applications of Graphs

1. (a) Suppose you have a container that is an inverted cone (similar to Figure 3.23(a)) which is being filled with water flowing at a constant rate. Sketch a graph of this event where the input is time and the output is the height of the water in the container. How does this differ from the graph in Example 3?

   (b) Repeat this exercise assuming your container is shaped as in Figure 3.23(b):

(a)   (b)

FIGURE **3.23**

2. Suppose a person left an office, walked slowly down a hall at a constant rate, stopped to talk to someone for a few seconds, and ran back to the office when the phone rang. Sketch a graph of this event where time is the input and the distance from the person to the office door is the output.

3. The graphs in Figure 3.24 were created using a calculator attached to a motion detector. The graphs use time (in seconds) as the input and the distance (in feet) the person was away from the motion detector as the output. For each of the following graphs, describe in words the motion of the person walking. Indicate if the person was walking away or towards the motion detector and if the person was speeding up or slowing down.

(a) 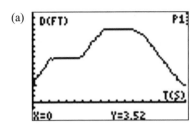   (b)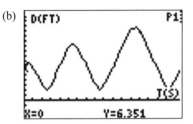

FIGURE 3.24

4. The graph in Figure 3.25 represents a child's fever where the input is time and the output is the child's temperature. When did the fever start to decline? How do you know?

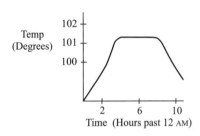

FIGURE 3.25

5. Sketch a plausible time-distance graph representing where you were during the 3 hours following this class the last time it met. Let t be the time from the end of this class to 3 hours later. Let $y = d(t)$ be the distance from the classroom. Write a couple of sentences explaining your graph.

6. Match each of the following function transformation operations with the correct graphical transformation.

    A. $y = f(x + c)$ (i)   vertical stretch
    B. $y = cf(x)$ (ii) horizontal shift
    C. $y = f(x) + c$ (iii) horizontal stretch
    D. $y = f(cx)$ (iv) vertical shift

7. Given the accompanying graph of $y = f(x)$, match the following function transformations to the graphs in Figure 3.26.

    (a) $y = f(2x)$

    (b) $y = 2f(x)$

    (c) $y = f(\frac{1}{2}x)$

    (d) $y = f(x) + 2$

    (e) $y = f(x - 2)$

    (f) $y = f(x + 2)$

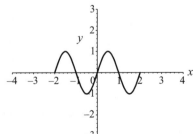

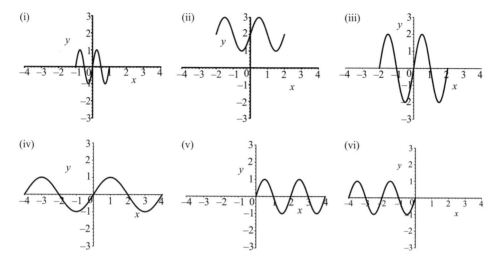

FIGURE 3.26

8. Suppose that $y = t(c)$ represents the function in which c is the number of credits you are taking this semester and $y = t(c)$ is the cost of tuition. If the cost per credit increases by 10% (which means the cost of tuition is 110% times last year's cost), what function represents the new cost of tuition?

    (a) $y = t(1.1c)$ (b) $y = t(c + 1.1)$ (c) $y = t(c) + 1.1$ (d) $y = 1.1t(c)$

9. Suppose that $y = t(c)$ represents the function in which c is the number of credits you are taking this semester and $t(c)$ is the cost of tuition. Suppose a change was made in the cost of tuition such that $y = 100 + t(c)$ describes the new tuition function. This new function describes which of the following situations?

    (a) The number of credit hours has increased by 100.

    (b) The cost per credit has increased by $100.

    (c) Tuition has increased by 100%.

    (d) The total cost of tuition has increased by $100 (e.g., a technology fee was added).

10. Which of the following type(s) of transformations will *always* result in all the points on the graph moving. (Circle all that apply.)

    (a) adding a constant to the output

    (b) multiplying the output by a constant

    (c) adding a constant to the input

    (d) multiplying the input by a constant

11. Let $y = f(x)$ as in Figure 3.27. Sketch a graph and give the **domain and range** for each of the following.

(a) $y = f(x) - 1$

(b) $y = f(x - 1)$

(c) $y = 2f(x)$

(d) $y = f(x) + 2$

(e) $y = \frac{1}{2}f(x)$

(f) $y = f\left(\frac{1}{2}x\right)$

(g) $y = f(2x)$

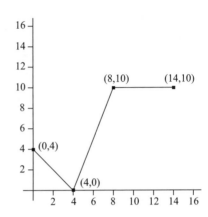

FIGURE 3.27

12. Let $y = w(x)$ be as given in Figure 3.12 in this section. Explain how to walk to produce each of the following:

   (a) $h(x) = w(x) - 2$

   (b) $j(x) = \frac{1}{3}w(x)$

   (c) $t(x) = w(x - 1)$

   (d) $s(x) = w(3x)$

13. Suppose a navy submarine was exercising tactics by beginning at sea level, diving to -300 feet in $\frac{1}{2}$ hour, then climbing to -100 feet in $\frac{1}{2}$ hour. It then dove to -500 feet in 1 hour where it rested for 2 hours. Let $y = s(t)$ be the function where t is time and $s(t)$ is the distance of the submarine from the surface.

   (a) Sketch a graph of $y = s(t)$. Let the horizontal axis represent sea level.

   (b) Sketch a graph of $y = s(\frac{1}{2}t)$. What did the submarine do differently for this graph?

14. The following graph is the original temperature function from Example 4. Suppose the homeowner wants the furnace to start warming up the house at 5 AM and the

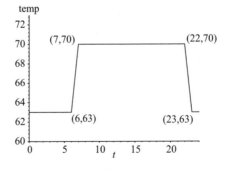

FIGURE 3.28

temperature to start decreasing at 9 PM. The homeowners want the high temperature in the house to still be 70° and the low temperature to still be 63°.

(a) Sketch this new scenario on the original graph.

(b) If $y = h(t)$ describes our original heating scenario, what function notation describes the graph you just drew?

Applications of Graphs: Activities and Class Exercises

1. **Height versus Weight.** For the first two years of a child's life, the function $w(h) = 1.4h - 20$ can be used to determine the average weight, in pounds, of a child whose height (or length) is given in inches.

(a) Find the value for $w(30)$ and describe what it means in terms of height and weight of a child.

(b) Suppose the weight was now calculated in kilograms instead of pounds and the height continued to be calculated in inches.

   i. Would this new function be a change in input or a change in output?

   ii. There are approximately 0.45 kilograms in 1 pound, so the new function could be described as $y = 0.45w(h)$. Find an equation for this new function.

   iii. What is the average weight, in kilograms, of a child that is 30 inches tall?

   iv. Graph $y = w(h)$ and $y = 0.45w(h)$ on the same set of axes in such a way that you can see the origin and both the x and y-axes on your graph. Describe the difference between the two graphs in terms of a vertical stretch (or compression). Be specific.

(c) Suppose the height was now calculated in centimeters instead of inches and the weight continued to be calculated in pounds.

   i. Would this new function be a change in input or a change in output?

   ii. There are 2.54 centimeters in 1 inch. Using this, find the average weight, in pounds, of a child that is 76.2 centimeters tall.

   iii. Explain why the new function $y = m(c)$ (where the input is in centimeters and the output is in pounds) is equal to $y = w(\frac{c}{2.54})$.

   iv. Find an equation for $y = m(c)$.

   v. Graph $y = w(h)$ and $y = m(c)$ on the same set of axes in such a way that you can see the origin and both the x and y-axes on your graph. Describe the difference between the two graphs in terms of a horizontal stretch (or compression). Be specific.

(d) Let's put both of these transformations together by finding the function in which the input is height in centimeters and the output is the weight in kilograms. In other words, find the formula for $y = 0.45w(h/2.54)$. [Hint: Do the input transformation first and then do the output transformation. As a check, take a point such as $(20, 8)$ that fits $w(h)$ and convert the 20 inches to centimeters (multiply by 2.54). Also, convert the 8 pounds to kilograms (multiply by 0.45). Make sure the new point fits your new function.]

2. **Race.** Runner Joe LeMay posts reports on the web describing races he participated in during 1998.[3] On July 20, 1998, Joe competed in the Lehigh River relay run, a race where a 5 person team has each person run a 5 mile leg. Joe recorded his mile "splits" (i.e., times recorded at each mile). Joe's mile splits are recorded in the following table.

Mile	1	2	3	4	5
Time	4:58	9:48	14:53	20:11	25:15

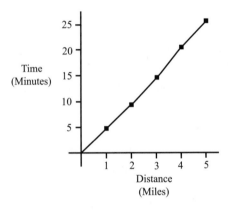

FIGURE 3.29

(a) Suppose Joe ran a 3 mile warm up before the race began.

  i. Would the 3 miles affect the input or output?

  ii. Sketch the graph that would result if Joe included this into his total miles.

  iii. If $y = d(t)$ was Joe's original function, what is the new function?

(b) Suppose Joe was behind several hundred runners at the start. By the time Joe reached the starting line and officially began his race, the clock had him timed at 10 seconds into the race.

  i. Would the 10 seconds affect the input or the output?

  ii. Sketch the graph that would result if Joe's starting time was clocked as 10 seconds into the race.

  iii. If $y = d(t)$ was Joe's original function, what is the new function?

(c) Joe would like to reduce his overall time by a factor of 1/50 (a 2% reduction).

---

[3] Joe LeMay's 1998 Races, 6 June 2002, <http://www.jlemoo.tripod.com/running/run1998.htm>.

    i. Would this reduction affect his input or output?

    ii. Sketch the graph that would result if Joe reduced his overall time by a factor of 1/50.

    iii. If $y = d(t)$ was Joe's original function, what is the new function?

(d) Suppose Joe was unsatisfied with his time for the last two miles of the race. He felt he should be able to run the last two miles 18 seconds faster while continuing to run the first 3 miles in 14:53 minutes.

    i. Sketch the graph that would result if he kept his original pace for miles 1 to 3, and then ran a pace that was 18 seconds faster for miles 4 and 5.

    ii. How long did Joe originally take to run the last 2 miles?

    iii. If he runs the last 2 miles 18 seconds faster, how long will it take Joe to run the last 2 miles?

    iv. Running the last two miles 18 seconds faster means new time=a(old time). What is a?

    v. If $y = d(t)$ was Joe's original function, what is the new function? [Hint: This will be a piecewise function.]

3. **Using the Motion Detector.** Produce each of the following graphs by having one (or more) of your group members walk in front of a CBR (motion detector). Include a description of how the person walked in order to obtain the appropriate graph. Indicate when the person walked toward or away from the motion detector as well as whether they were walking fast or slow.

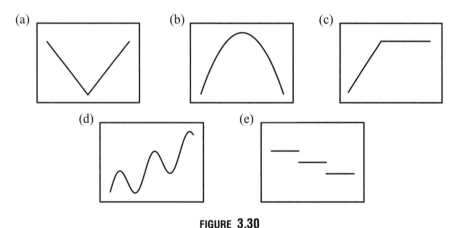

FIGURE 3.30

4. **Cost of DVD Players.** DVDs were first introduced into the media market in 1997. The rate at which DVDs have become household fixtures is far greater than the rate at which VCRs were first assimilated. A DVD player cost $600 in January of 1998 and dropped in price to $177 in January of 2002. The following graph displays the price trend for DVD players from 1998 to 2002.[4]

---

[4] "Taking Command." *U.S. News and World Report* 11 Mar 2002: 44.

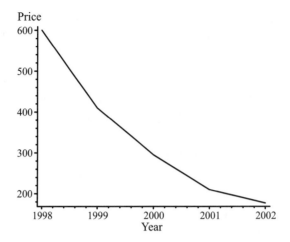

FIGURE 3.31

(a) Is the graph increasing or decreasing? Is it concave up or down? What does this tell you about the price of DVD players?

(b) Looking at the graph what do you think the price of a DVD player will be in 2005? Briefly justify your answer.

(c) Suppose DVD players started out at a price of $600 in 2000 rather than 1998.

    i. Assuming the original change in price remains the same, sketch the new graph.

    ii. If the original function is $y = p(t)$, what is your new function?

(d) Suppose DVD players started out in 1998 at the higher price of $800.

    i. Assuming the original change in price remains the same, sketch the new graph.

    ii. If the original function is $y = p(t)$, what is the new function?

(e) Suppose both things in parts (c) and (d) occurred, and DVD players started out in 2000 at a price of $800.

    i. Assuming the original rate of change of price remains the same, sketch the new graph.

    ii. If the original function is $y = p(t)$, what is the new function?

(f) Suppose advertising and a larger acceptance of DVD players occured at a greater rate and caused the price of DVDs to decrease 1.2 times faster.

    i. Assuming DVDs still started out in 1998 at a price of $600, sketch the new graph.

    ii. If the original function is $y = p(t)$, where t is measured in years since 1998, what is the new function?

    iii. Why is it necessary to use $t = 0$ for 1998 for part (ii)?

4

Displaying Data

When data is displayed, it is important that the information it conveys can be easily understood. Sometimes data is displayed as a table of numbers and sometimes it is displayed as a graph. In this section, we will look at what types of graphs and tables are most appropriate for displaying various types of data.

Data Involving One Variable

The data set listed in Table 4.1 consists of the heights, in inches, of fifty women. This data involves just one variable, height. Data involving only one variable is the simplest type of data. The data in Table 4.1 is listed in the order it was collected. Looking through the numbers, we notice that most of the heights are in the sixties with a few in the seventies. Upon a closer inspection, we discover that the smallest height is 60 and the largest is 72. It is difficult, however, to get a feeling for the average height of this sample just by scanning the numbers in Table 4.1. It is also difficult to get a sense as to what portion of the data is close to the average.

Frequency Distributions

We need to organize the data differently to gain a better understanding of these heights. One common method of organizing data is a frequency distribution. A **frequency**

TABLE 4.1
Heights (in inches) of 50 women.

67	64	65	65	64	65	67	65	65	65
60	67	67	72	65	63	67	66	66	70
64	62	66	67	66	66	64	67	64	61
60	70	68	61	64	65	66	69	71	63
60	68	65	66	62	68	71	68	64	68

TABLE **4.2**

A frequency distribution for the heights of 50 women.

Height	60	61	62	63	64	65	66	67	68	69	70	71	72
Frequency	3	3	1	2	7	9	6	7	5	1	2	2	1

distribution is a table in which the data is grouped into categories and the number of pieces of data in each of these categories is given. Since the numbers in Table 4.1 are integers from 60 to 72, our frequency distribution will consist of thirteen categories: $60, 61, 62, \ldots, 72$.[1] Counting the number of heights in each category and listing that number as its frequency, we obtain the frequency distribution shown in Table 4.2.

The frequency distribution gives us a better understanding of how the heights are distributed. For example, we can easily see that the heights range from 60 inches to 72 inches. We also see that most of the heights are between 64 and 68 inches. From this, we might guess that the average height of this group of fifty women is close to 66 inches.

Histograms

A **histogram** is a graphical representation of a frequency distribution. In a histogram, the values are divided into equal intervals (shown on the horizontal axis) and the *frequency* for each interval is displayed by forming rectangles whose heights are the frequencies. This gives us a visual representation of the number of data values in each category. The advantage of a histogram is that, with just a quick glance, we see how the data is distributed. A histogram of our height data is shown in Figure 4.1.

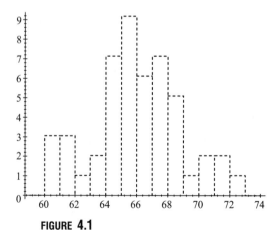

FIGURE **4.1**

A histogram for the heights of 50 women.

---

[1] There are other ways to group this data. For example, we could let our categories consist of 60 and 61, 62 and 63, ..., 72 and 73.

TABLE 4.3
Heights and arm spans (in centimeters) of 16 students.

Height	Arm Span	Height	Arm Span
152	159	173	170
156	155	173	169
160	160	173	176
163	166	179	183
165	163	180	175
168	176	182	181
168	164	183	188
173	171	193	188

Data Involving Two Variables

Scatterplots

We are often interested in relationships between two variables. For example, you may have heard that your height is about the same as the distance from finger tip to finger tip of your outstretched arms (or arm span). To investigate this relationship, we collected data from 16 students. This data is shown in Table 4.3.

Looking through these data, notice that most everyone had an arm span that is close to his or her height. In some cases, the arm span is greater than the person's height (such as the first person listed). Other people have heights greater than their arm span (such as the last person listed). It is easier to see these relationships if we plot the data points. Since the data consists of two variables, we can think of the table as a listing of ordered pairs or points. A **scatterplot** is a graph used to plot data as ordered pairs. When constructing a scatterplot, it is important to put the independent (or input) variable on the horizontal axis and the dependent (or output) variable on the vertical axis. In this example, many would think that arm span depends on height. In that case, we would have height as our independent variable and put it on the horizontal axis and arm span as the dependent variable and put it on the vertical axis.[2] Figure 4.2 is a scatterplot of the data from Table 4.3. From this scatterplot, you can see that the points are generally increasing. This is an example of a positive association. Two variables have a **positive association** when an *increase* in the value of one variable leads to an *increase* in the value of the other variable. In our situation, this means that taller students tend to have longer arm spans.

---

[2] Some might argue that height does not depend on arm span nor does arm span depend on height. This does not mean that height and arm span are not related. It just means that one is not the cause of the other nor is one the response to the other. In cases like this, it does not matter which variable is on the horizontal axis and which one is on the vertical.

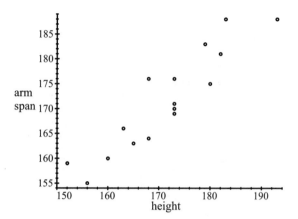

FIGURE **4.2**
A scatterplot of students' heights and arm spans.

A negative association would occur if the opposite were true. Two variables have a **negative association** when an *increase* in the value of one variable leads to a *decrease* in the value of the other variable. An example of this is the relationship between latitude and average January temperature for cities in the United States. As the latitude increases (moving farther north), the average January temperature decreases. Figure 4.3 graphs the latitude and average January temperature for twelve cities in the eastern United States.[3]

The data in Figure 4.2 and Figure 4.3 have linear patterns although the points in each do not all lie on a line. For Figure 4.2, our original assumption that a person's height was about the same as his or her arm span can be tested by superimposing the line $y = x$ onto our scatterplot. (See Figure 4.4.) The input and output are exactly the same for each point on the line $y = x$. Therefore, it serves as a reference for everyone whose height is exactly the same as the length of their arm span. This line does a fairly good job of

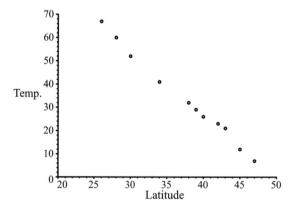

FIGURE **4.3**
A scatterplot of latitude and average January temperatures for cities in the eastern United States.

---

[3] The data was gathered from *The World Almanac and Book of Facts 1996*, pp 180, 594, 595.

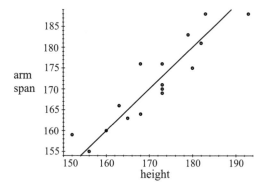

FIGURE **4.4**
A scatterplot of students' heights and arm spans with the line $y = x$.

approximating our data. While only one point, $(160, 160)$, lies on the line, all the points are close so we can conclude that our assumption is basically true. The points that lie above the line have larger outputs than inputs. Therefore, they represent people with arm spans that are longer than their height. The opposite is true of those people whose points lie below the line.

xy-Lines

Sometimes the data we want to represent graphically is a function. This is often true when one of the variables is time. As an example, consider the change in a monthly natural gas bill for a home over time. For any given month (the input), there is only one monthly cost (the output), so this is a function. Often, with data that is dependent on time, we are interested in finding trends. We could plot this data as a scatterplot as we did with the arm span and height data. However, when you are looking for trends, an xy-line tends to give a better display. An xy-**line** or **trend line** is a scatterplot that represents a function where the consecutive points are connected with line segments. Figure 4.5 shows an xy-line where the input is time in months (1 = May 1993, 2 = June 1993, etc.) and the output is the monthly cost in dollars of natural gas. By connecting the points with line

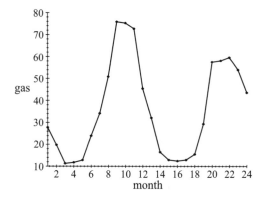

FIGURE **4.5**
An xy-line of monthly gas costs.

segments, we can easily see how the monthly cost of natural gas fluctuates throughout the year. You have probably seen graphs like this in newspapers showing how the stock market or the price of an item changes over time, for example.

Application: House Prices

We discussed how we can detect relationships between two variables. Let's see how this applies to looking at house prices versus square footage. The cost of a new house is partly determined by its total footage. Table 4.4 compares the area and cost of 14 relatively new homes built in western Michigan in 1998.[4]

TABLE **4.4**
Area and Price of some new homes in western Michigan in 1998.

Area	1345	1920	1729	2100	1657	1431	3344
Price (ft^2)	$94,500	$157,900	$199,900	$229,900	$159,900	$105,900	$204,500
Area	2332	1317	2611	1610	2000	971	1678
Price (ft^2)	$225,500	$130,900	$229,000	$174,900	$219,900	$107,900	$163,900

Just looking at the table, it is difficult to tell if there is any relationship between cost and square footage. A scatterplot representing the data from the table is shown in Figure 4.6. The area is on the horizontal axis and the price is on the vertical axis because cost depends on area. The data seem to exhibit a somewhat linear relationship.

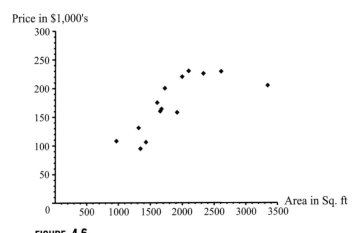

FIGURE **4.6**
A scatterplot comparing square footage to cost for 14 houses.

---

4 *Woodland Realty*, 22 November 1998, <http://www.woodlandrealty.com>.

Example 1.   According to the Census Bureau, the average house built in the U.S. in 1996 cost $59 per square foot and the average lot cost was $41,550.[5] Using these two parameters, write the symbolic form of the function that you could use to estimate the cost of a new house and lot built in the United States in 1996, given the area of the home.

Solution.   Building a new house involves finding the cost of the lot and the cost of building the house. So we know that the equation involves the sum of these two values. We know the average cost of the lot is $41,550 and we also know that the cost of building a house is $59 **per square foot**. This means we have to multiply the square footage of the house by the cost per square foot. Doing this operation and adding our product to the lot price, we get the equation

$$C = 59a + 41,550,$$

where a is the area of the house in square feet and C is the total cost of the house and the lot. ∎

The line determined in Example 1 is shown in Figure 4.7, along with our original scatterplot from Figure 4.6. Notice that most of the points are above this line. This means that many of the houses built in 1998 in western Michigan were more expensive than the U.S. average in 1996. Only three points lie below this line.

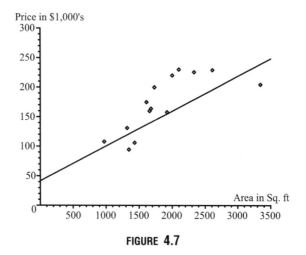

FIGURE **4.7**

Reading Questions for Displaying Data

1. How are a frequency distribution and a histogram similar? How are they different?

2. What are the advantages of displaying data in a frequency distribution?

3. What are the advantages of displaying data in a histogram?

---

[5] *The Holland Sentinel* 23 November 1997, p. B1.

4. The first rectangle in the histogram in Figure 4.1 represents the 3 heights that were 60 inches. The base of this rectangle extends from 60 to 61 on the horizontal axis. Technically this means that the heights contained in that interval could be anything greater than or equal to 60 and less than 61. Because the data is in integer units, this interval will consist of only heights that were exactly 60 inches. Looking at just the histogram, how many heights were 62 inches? How many were 66 inches? Check your answers with Table 4.2.

5. Using your calculator or software:

    (a) Construct a histogram similar to Figure 4.1 using the numbers in Table 4.2. Make a sketch of the histogram.

    (b) If you wanted to group the data in categories of three heights combined (i.e., 60 to 62, 63 to 65,..., 69 to 71, 72 to 74), how would you do this on your calculator or software?

6. Answer the following questions about histograms using the data given below.

$$1 \quad 1 \quad 2 \quad 2 \quad 4 \quad 5 \quad 5 \quad 8 \quad 9 \quad 12 \quad 12 \quad 12$$

    (a) Construct a histogram by hand using three groups: 1 to 4, 5 to 8, 9 to 12.

    (b) Construct the same histogram using your calculator or software.

    (c) With your calculator or software, construct a histogram using four groups: 1 to 3, 4 to 6, 7 to 9, 10 to 12. Make a sketch of the histogram.

    (d) Why do your two histograms look different? What does this tell you about constructing histograms with different groupings?

7. Use the two histograms in Figure 4.8 to do the following:

    (a) Estimate the average for each of the two histograms.

    (b) The same data was used to create both histograms. For which one was it easier to estimate the average?

    (c) Based on your answer to part (b), what "rule of thumb" should you use when drawing histograms?

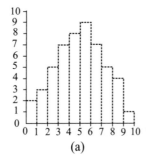
(a)

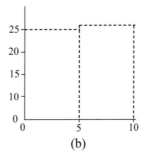
(b)

FIGURE **4.8**

8. Why does the scatterplot of people's heights and arm spans in Figure 4.2 have a positive association?

9. For each of the following, determine if the two variables would have a positive association, a negative association, or no association. Briefly explain.

   (a) size of a building and cost of heating or cooling

   (b) automobile's age and gas mileage

   (c) size of a city and number of violent crimes

   (d) number of songs on a CD and cost of the CD

   (e) amount of time spent watching TV and amount of time spent studying in the library

10. If you were to graph each of the following on a scatterplot, which variable would go on the horizontal axis or does it not matter? Explain your answer.

    (a) size of a building and cost of heating or cooling

    (b) automobile's weight and gas mileage

    (c) size of a city and number of violent crimes

    (d) car's price and age

    (e) number of years of education past high school and height

11. What can we say about the height and arm span of people whose points lie above the line $y = x$ in Figure 4.4? What can we say about the height and arm span of people whose points lie below the line $y = x$ in Figure 4.4?

12. How is an xy-line different from a scatterplot? What type of data is usually plotted using an xy-line?

13. Using your calculator or software and the height and arm span data given in Table 4.3, make a scatterplot of the data.

14. Graph the given data sets as both a scatterplot and an xy-line. Sketch both graphs for each data set. Which is the better graph? Why?

    (a)

x	1	2	3	4	5	6	7	8	9	10	11	12	13	14	15	16
y	27	30	36	29	20	22	18	14	21	22	35	34	32	28	28	19

    (b)

x	1200	1300	1100	1250	1600	1500
y	23.5	35.6	25.2	33	48.9	37.2

15. The graph in Figure 4.9 represents weights, in grams, of individual pieces of Starburst candies.

    (a) What type of graph is this?

    (b) How many Starbursts weigh between 5.0 and 5.2 grams?

    (c) Just by looking at the graph, give the approximate average weight of an individual Starburst candy.

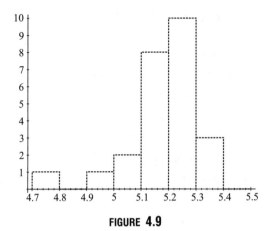

FIGURE **4.9**

16. Use an appropriate graph from this section to display the following data. Make a sketch of each graph.

    (a) This data set represents weights (in kilograms) of 20 young women.

    $$52 \quad 75 \quad 63 \quad 60 \quad 63 \quad 65 \quad 47 \quad 62 \quad 35 \quad 55$$
    $$55 \quad 51 \quad 72 \quad 58 \quad 53 \quad 53 \quad 65 \quad 50 \quad 50 \quad 71$$

    (b) The following table shows a listing of the area of some homes that were for sale in Colorado Springs in 2002.[6]

Area (sq. ft.)	Price	Area (sq. ft.)	Price
1668	$149,000	2687	$189,950
2342	$184,900	1856	$149,900
2967	$250,000	2409	$234,000
1507	$155,000	2410	$187,500
2050	$157,000		

Displaying Data: Activities and Class Exercises

1. **Gasoline Prices.** Gasoline is produced by processing crude oil. In this activity, we are interested in looking at the relationship between gasoline prices and crude oil prices.

---

[6] *Colorado Springs Real Estate Listings*, 17 June 2002, <http://www.colorado-springs-real-estate-listings.com /real-estate-listings.html>.

(a) Since gasoline is produced from crude oil, do you think that there should be a positive relationship between the cost of a gallon of crude oil and the cost of a gallon of gasoline?

(b) The following histogram gives the average cost of a gallon of crude along with the average cost of a gallon of gasoline. A graph like this was once used to try to convince people that there was a positive relationship between the costs of oil and gasoline. Looking at the histogram, what do you conclude about this relationship?

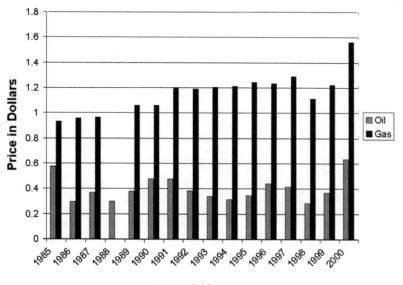

FIGURE **4.10**

(c) The data in Table 4.5 shows the price of a gallon of gasoline and the price of a gallon of crude oil from 1985 to 2000.[7] The previous histogram was based on the data from Table 4.5. Enter the price of gasoline into List 1 on your calculator and the price of oil (for the corresponding year) into List 2. Construct a scatterplot and sketch it.

(d) Looking at your scatterplot, do the points show a positive association, negative association, or no association?

(e) What does your answer to part (d) tell you about the relationship between the price of a gallon of gasoline and the price of a gallon of oil?

(f) Is your answer to part (b) different than your answer to part (e)? Explain.

(g) Which graph, the histogram or scatterplot, better gives the actual relationship between the costs of crude oil and gasoline? Explain.

---

[7] "U.S. Average Gasoline Prices." *Louisiana Mid-Continent Oil and Gas Association*, 25 June 2002, <http://www.lmoga.com/gasoline.html>.

TABLE **4.5**

Year	Average Price of Gasoline ($ per gallon)	Average Price of Oil ($ per gallon)
1985	0.931	0.574
1986	0.957	0.298
1987	0.964	0.367
1988	NA	0.299
1989	1.059	0.378
1990	1.059	0.477
1991	1.196	0.477
1992	1.190	0.383
1993	1.205	0.339
1994	1.212	0.314
1995	1.242	0.348
1996	1.231	0.440
1997	1.291	0.417
1998	1.115	0.287
1999	1.221	0.370
2000	1.563	0.633

2. **Heights of Couples.** The graph in Figure 4.11 represents the heights, in inches, of 24 wives and husbands.

   (a) What type of graph is this?

   (b) What are the heights of wives whose husbands are 73 inches tall?

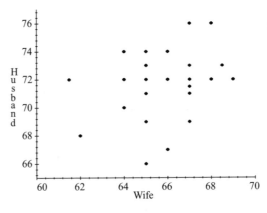

FIGURE **4.11**

(c) Each wife has a husband that is taller. What are the heights of the wife and husband whose heights are closest together? What are the heights of the wife and husband whose heights are farthest apart?

3. **Electric Bill.** The graph in Figure 4.12 represents the monthly cost of electricity for a house. Time, in months, is on the horizontal axis where 1 represents May, 2 represents June, 3 represents July, and so on for a two year period.

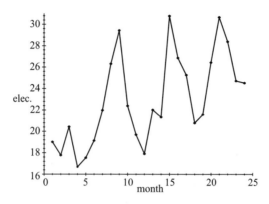

FIGURE 4.12

(a) What type of graph is this?

(b) What time interval is represented by the graph? Give your answer in terms of the first month and year and the last month and year.

(c) What was the cost of electricity each September?

(d) During what time of year is the most electricity used? During what time of year is the least electricity used? Why do you think this occurs?

4. **Fifth-Third Bank Run**. The Fifth-Third River Bank Run is an event composed of a 5k race and a 25k race held in downtown Grand Rapids, MI. It has been held on the second Saturday of May for the past twenty five years. The 25k race is the event's showcase with over 4,700 athletes competing in the event. The following two charts give some of the results of the Fifth-Third River Bank Run in 2002.[8] Use this data to answer the following questions.

(a) Calculate the percentages in each category for Men ages 20–24 and for Men ages 50–54.

(b) Make a histogram for Men ages 20–24 using the *percentages* rather than the number of runners. Do the same for Men ages 50–54.

(c) What is similar about the two histograms? What is different?

(d) Explain the similarities and the differences in terms of the ages of the runners.

---

[8] *Classical Race Management*, 17 June 2002, <http://www.classicrace.com/results/02/25k.htm>.

25K-Men Ages 20–24

Time Intervals	# Runners	% Total
1:00:00-1:15:00	1	
1:15:01-1:30:00	16	
1:30:01-1:45:00	23	
1:45:01-2:00:00	40	
2:00:01-2:15:00	38	
2:15:01-2:30:00	23	
2:30:01-2:45:00	13	
2:45:01-3:00:00	0	
3:00:01-3:15:00	0	
3:15:01-3:30:00	0	
Total	154	

25K-Men Ages 50–54

Time Intervals	# Runners	% Total
1:00:00-1:15:00	0	
1:15:01-1:30:00	1	
1:30:01-1:45:00	7	
1:45:01-2:00:00	59	
2:00:01-2:15:00	90	
2:15:01-2:30:00	80	
2:30:01-2:45:00	45	
2:45:01-3:00:00	20	
3:00:01-3:15:00	3	
3:15:01-3:30:00	1	
Total	306	

5. **Textbook Prices.** Going to the college bookstore can be frustrating. There are long lines, and people are always hurrying to get the used books first. The most depressing part is often paying the bill at the end. To see whether or not there is a relationship between the number of pages in a textbook and the cost, 12 books were selected at random from a web site selling college textbooks.[9]

# of Pages	Price
1185	$122.95
766	$105.00
1116	$79.95
1300	$117.95
689	$78.75
701	$81.33

# of Pages	Price
480	$55.00
993	$124.75
2112	$74.95
1120	$130.95
1616	$102.00
1175	$81.20

(a) Sketch a scatterplot of the data. What, if any, is the relationship between price and the number of pages in a textbook?

(b) This data was collected using textbooks from a variety of different subject areas. Do you think this has an effect on the relationship in your scatterplot?

(c) If a sample were to be taken using only math textbooks, would you expect there to be a positive association, negative association, or no association between the number of pages and the price of the textbook? Explain.

6. **Basketball Salaries.** The table below lists the teams in the National Basketball Association along with their total team payrolls in millions of dollars and the winning

---

[9] *Barnes and Noble.com College Textbooks Homepage*, 24 June 2002, <http://www.barnesandnoble.com/textbooks/index.asp?sit=T&>.

percentage for the 2001–2002 season.[10] We want to determine if the team owners are getting their money's worth from their highly paid players.

Team	Payroll (in millions)	01–02 Winning Percentage
New York	$85.25	0.366
Portland	$84.00	0.598
Philadelphia	$57.79	0.524
Dallas	$56.51	0.695
Milwaukee	$55.57	0.5
Sacramento	$54.76	0.744
Phoenix	$54.70	0.439
Minnesota	$54.51	0.610
Utah	$54.33	0.537
Denver	$53.69	0.329
New Jersey	$53.60	0.634
Indiana	$53.21	0.512
L.A. Lakers	$52.99	0.707
Miami	$52.92	0.439
Toronto	$52.66	0.512
Washington	$51.07	0.451
Memphis	$50.29	0.280
Charlotte	$49.73	0.537
Atlanta	$49.33	0.402
Houston	$49.24	0.341
Orlando	$48.32	0.537
Boston	$47.47	0.598
Golden State	$47.17	0.256
San Antonio	$46.49	0.707
Cleveland	$45.92	0.354
Seattle	$44.99	0.549
Chicago	$42.54	0.256
Detroit	$41.71	0.610
L.A. Clippers	$33.73	0.476

(a) Looking through the table, does it seem as though the higher paid teams had higher winning percentages? If so, were there exceptions to this rule?

---

[10] *USA Today*, 12 June 2002, <http://www.usatoday.com/sports/nba/stories/2001-2002-salaries-was.htm>.

(b) Make a scatterplot of this data with the team payroll on the horizontal axis and the winning percentage on the vertical axis.

(c) Does your scatterplot have a positive or a negative association?

(d) Which team would you say is the best value for the money? Why?

(e) Which team would you say is the worst value for the money? Why?

7. **Basketball Players' Statistics.** Two statistics kept of a basketball player's performances are average points per game (the average number of points a player scores each game) and average assists per game (the average number of passes a player makes to another player who scores after receiving the pass). In order to see if there is a relationship between average points per game and average assists per game, we selected 20 NBA players' statistics from the 2001–2002 season.[11]

Player	Average Points per Game	Average Assists Per Game
Iverson	31.4	5.5
Malone	22.4	4.3
Miller	16.5	10.9
Nash	17.9	7.7
O'Neal	27.2	3.0
Nowitzki	23.4	2.4
Pierce	26.1	3.2
Tinsley	9.4	8.1
Davis	18.1	8.5
Rose	23.8	5.3
McGrady	25.6	5.3
Webber	24.5	4.8
Kidd	14.7	9.9
Williams	14.8	8.0
Payton	22.1	9.0
Duncan	25.5	3.7
Bryant	25.2	5.5
Van Exel	21.4	8.1
Stockton	13.4	8.2
Marbury	20.4	8.1

(a) Do you think there is a negative, positive, or no association between points per game and assists per game? Why?

---

[11] *SportingNews.com-NBA*, 11 June 2002, <http://www.sportingnews.com/nba/statistics>.

(b) Sketch a scatterplot based on the data in the above table.

(c) Looking at the scatterplot, what type of association do you observe?

(d) Using your answer to part (c) as justification, complete the following sentence, "As a basketball player's points per game increases, his assists per game _____."

(e) Name a player who does not seem to follow the conclusion you gave in part (d).

5

Describing Data: Mean, Median, and Standard Deviation

Displaying data in the form of tables and graphs gives an overall picture of the data's behavior. It is also common, however, to describe a data set using numerical descriptors. Numbers make it easy to compare two or more data sets. In this section, we will describe two methods of measuring the center and the spread of a data set. We will compare these methods and discuss how to decide which one is the most appropriate measure of center for a given data set.

Measuring Center

The most common way of measuring the center of a set of data is the ordinary arithmetic average, also known as the mean. To determine the **mean** of a data set, you add up the data values and divide by the number of pieces of data. For example, to find the mean of $\{2, 6, 4, 5, 9\}$, you find the sum of the numbers, $2 + 6 + 4 + 5 + 9 = 26$, and divide by five since there are five numbers in this set. For this example, the mean is $\frac{26}{5} = 5.2$. We write the mean as $\overline{x}$, which is usually pronounced "x-bar."

You can think of the mean as a way of "levelling off" a data set. For example, our data set $\{2, 6, 4, 5, 9\}$ has the same total as the data set $\{5.2, 5.2, 5.2, 5.2, 5.2\}$. Therefore, the mean can be thought of as the amount every data point receives if resources (the total) are divided equally.

The procedure for finding the mean can be written using mathematical symbols. If there are n numbers, denoted by the generic names $x_1, x_2, x_3, \ldots, x_n$, their mean is

$$\overline{x} = \frac{x_1 + x_2 + x_3 + \cdots + x_n}{n}.$$

This can also be expressed as

$$\overline{x} = \frac{\sum_{i=1}^{n} x_i}{n}.$$

The Σ is the upper case Greek letter sigma, used as a summation symbol. The expression, Σx_i, is just a symbolic way of saying "add up the numbers," $x_1 + x_2 + x_3 + \cdots + x_n$.

Another common way of measuring the center of a set of data is the median. The **median** is the middle number of an *ordered* set of data. To determine a median for the data set $\{2, 6, 4, 5, 9\}$, we first order the numbers from smallest to largest[1]: $2, 4, 5, 6, 9$. The middle number is the median. In this case, the median is 5. If you are trying to find the median for an even number of values, the median is the average of the middle two in the ordered set. For example, to determine the median for the data set $\{2, 1, 8, 5, 9, 4\}$, we first order the numbers from smallest to largest as we did before: $1, 2, 4, 5, 8, 9$. Since there are an even number of data points, we take the average of the two middle data points, 4 and 5. In this case, the median is 4.5. The median of a data set has the property that the number of data points larger than the median is equal to the number of data points smaller than the median. This is why the median is a measure of center for the data set.

Example 1.   Barry Bonds has been playing major league baseball since 1986. In 2001, he hit a major league record of 73 home runs. The following table is the number of home runs he hit each year from 1986 to 2004.[2]

Year	'86	'87	'88	'89	'90	'91	'92	'93	'94	'95	'96	'97	'98	'99	'00	'01	'02	'03	'04
Home Runs	16	25	24	19	33	25	34	46	37	33	42	40	37	34	49	73	46	45	45

Determine the mean and median number of home runs hit by Bonds from 1986 to 2002.

Solution.   The mean number of home runs for Bonds is

$$\bar{x} = \frac{16+25+24+19+33+25+34+46+37+33+42++40+37+34+49+73+46+45+45}{19}$$

$$\approx 37.$$

This means that, on average, Barry Bonds hit 37 home runs each year. To determine the median number of home runs, we first order the data and then find the middle number.

$$16, 19, 24, 25, 25, 33, 33, 34, 34, \mathbf{37}, 37, 40, 42, 45, 45, 46, 46, 49, 73$$

Since this is an odd numbered data set, the median is just the middle number, 37. This tells us that there were approximately as many years when Bonds hit 37 or more home runs as there were years when he hit 37 or fewer home runs. ∎

Comparing the Mean and Median

Both the mean and median are commonly used measures of center. Often, these two measures give similar results. For example, we found earlier that the mean of $\{2, 6, 4, 5, 9\}$ is 5.2 while the median is 5. In Example 1, we saw that the mean number of annual home runs for Barry Bonds was exactly the same as the median, 37. However, there are times

---

[1] Alternatively, we could also have ordered the data from largest to smallest.
[2] *Barry Bonds*, 4 Oct. 2004, <http://sports.espn.go.com/mlb/players/stats?statsId=3918>.

when the median and mean are quite different. For example, a statistics class was asked to complete a questionnaire where one of the questions was, "How many hours per week do you watch television?" Some of the answers to that question were $\{3, 5, 50, 3, 5, 5, 1, 4\}$. The mean of this sample is 9.5 hours and the median is 4.5 hours. Notice that the mean is actually larger than all but one data point. The reason the mean and median are noticeably different is because of the impact of the data point 50. This data point is an outlier.

In general, an **outlier** is a data point that falls well outside of the overall pattern of the data. When calculating the mean, every number is added to the sum. Thus, 50 has a significant affect on the sum and therefore on the mean of this set. The median, however, is not impacted by outliers. The only thing that matters when calculating the median is the middle value of the ordered list. When finding the median, the 50 is considered simply as a number larger than the median, similar to an 8 or 9. For the data set involving watching television, the median provides a better sense of the center than does the mean.

When a distribution is symmetric, such as in Figure 5.1, the mean and median both occur near the center of the diagram and will be approximately equal. The median always occurs at the point that divides the histogram into two parts of equal area. We will represent the median in a histogram with the letter M. The mean always occurs at the point where the distribution would balance if it were made of a solid material. We will represent the mean in a histogram with the symbol Δ. For the distribution shown in Figure 5.1, both the mean and median are 25.

How do the mean and median compare if a distribution is not symmetric? Let's look at the following data set:

$$\{5, 5, 5, 5, 5, 10, 10, 10, 10, 10, 10, 15, 15, 15, 15, 15,$$

$$20, 20, 20, 20, 25, 25, 25, 30, 30, 30, 35, 35, 40, 45\}$$

This distribution clearly is not symmetric (see Figure 5.2). The mean (Δ) of this data set is approximately 18.7 and the median (M) is 15. Notice that the mean is larger than the median. The mean was "pulled" in the direction of the outliers (40 and 45). The histogram for this data set (shown in Figure 5.2) shows that this distribution is skewed

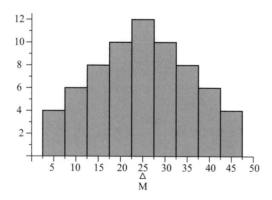

FIGURE 5.1
The mean and median are equal when the distribution is symmetric. In this distribution, they are both 25.

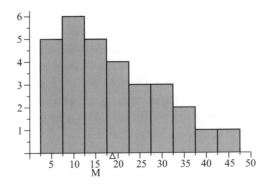

FIGURE 5.2
The mean of a distribution that is skewed to the right is larger than the median. In this distribution, the mean is approximately 18.7 and the median is 15.

to the right. A histogram is *skewed to the right* if the right-hand "tail" is longer than the left-hand "tail." The mean will be larger than the median when a distribution is skewed to the right.

Figure 5.3 shows the histogram of a distribution which is skewed to the left. For Figure 5.3, the mean (Δ) is approximately 6.2 and the median (M) is approximately 7. When a distribution is skewed to the left, the mean is smaller than the median. Once again, the mean was pulled in the direction of the outliers.

In general, how do you decide if the mean or the median is a better measure of center? The answer depends on what you are looking for and how the data are distributed. If the distribution is approximately symmetric (meaning there are no significant outliers), the mean and the median will be approximately the same. If the data are highly skewed (meaning there are significant outliers), then the median usually gives a better measure of center. However, there is information inherent in the computation of the mean that is not inherent in the median. The median only gives the middle of the data set. The mean gives the value of each data point assuming they were all exactly the same value.

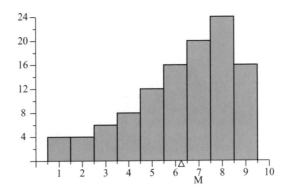

FIGURE 5.3
The mean of a distribution that is skewed to the left is smaller than the median. In this distribution, the mean is approximately 6.2 and the median is approximately 7.

For example, suppose a company has 100 employees and the mean salary is $28,000 and the median salary is $31,000. The total amount of money paid out in salaries is $100 \times \$28,000 = \$2,800,000$. You cannot deduce this type of information using the median.

Measuring Spread

Suppose a teacher has two small sections of the same course. On a particular test, Section *A* had scores of

$$50, 70, 70, 75, 80, 80, 100.$$

On the same test, Section *B* had scores of

$$50, 50, 50, 75, 100, 100, 100.$$

The teacher wants to describe these two sets of scores numerically. The teacher finds that Section *A* had a mean of 75 and a median of 75 and Section *B* also had a mean of 75 and a median of 75. However, even though the two sets of scores have the same mean and median, the distribution is quite different. To describe this difference, we need a measure representing the spread of the data. The simplest measure of spread is range. The **range** of a set of numbers is the smallest number subtracted from the largest number. In the case of the test scores, the range is $100 - 50 = 50$ for both Section *A* and Section *B*. The range is not sufficient, in this example, to distinguish between the two sets of scores.

Because the scores for Section *B* are clustered differently from those for Section *A* (see Figure 5.4), we need a measure of spread that takes into account more than just the highest and lowest numbers. One such measure is the standard deviation, which is the most common measure of the spread of a data set. Standard deviation can be thought of roughly as the average distance between the individual data points and the mean of the data set. The larger the standard deviation, the greater the spread of the data. To define standard deviation, we will first define an associated value known as variance. The **variance** of a set of numbers is the mean of the squared differences between each number of the set and the mean of the set. **Standard deviation** is the square root of the variance. To find the standard deviation of a data set by hand, do the following calculations:

1. Find the mean, $\overline{x}$, of the data set.
2. For each data point, x_i, in the data set, find the difference between it and the mean, $x_i - \overline{x}$.

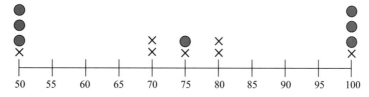

FIGURE 5.4

The ×s represent section A's scores, while the circles represent section B's scores.

3. Square the differences from step 2, $(x_i - \overline{x})^2$. [Note: Since the differences are squared, there is no distinction between data points below the mean and data points above the mean.]

4. Find the sum of the squared differences, $\Sigma(x_i - \overline{x})^2$.

5. Divide the sum of the squared differences by the number of data points in the set, $\Sigma(x_i - \overline{x})^2/n$. This is the variance.

6. Find the square root of the variance, $\sqrt{\frac{\Sigma(x_i - \overline{x})^2}{n}}$. This is the standard deviation.[3]

The lower case Greek letter sigma, σ, is often used to represent standard deviation. So

$$\sigma = \sqrt{\frac{\Sigma(x_i - \overline{x})^2}{n}}.$$

We will go through the process of computing the standard deviations for the two data sets in our sample. Typically, however, this process is done using a calculator or software.

Example 2.   Compute the standard deviation of the test scores for Section A,

$$50, 70, 70, 75, 80, 80, 100$$

and Section B

$$50, 50, 50, 75, 100, 100, 100.$$

Solution.   We mentioned earlier that the mean, $\overline{x}$, for both sets of scores is 75. We found the sum of the squared differences in the right-most entry of the last row in the following table.

Section A

	x_i	$x_i - \overline{x}$	$(x_i - \overline{x})^2$
	50	$50 - 75 = -25$	$(-25)^2 = 625$
	70	$70 - 75 = -5$	$(-5)^2 = 25$
	70	$70 - 75 = -5$	$(-5)^2 = 25$
	75	$75 - 75 = 0$	$(0)^2 = 0$
	80	$80 - 75 = 5$	$(5)^2 = 25$
	80	$80 - 75 = 5$	$(5)^2 = 25$
	100	$100 - 75 = 25$	$(25)^2 = 625$
$\Sigma(x_i - \overline{x})^2$			1350

So,

$$\sigma = \sqrt{\frac{\Sigma(x_i - \overline{x})^2}{n}} = \sqrt{\frac{1350}{7}} \approx 13.9$$

---

[3] The formulas that we will use for variance and standard deviation have an n in the denominator. This gives the standard deviation for a population. There are other formulas for sample variance and standard deviation that have $n - 1$ in the denominator and give the standard deviation for a sample.

for Section A. We can think of 13.9 as being roughly the average distance of the scores from the mean.

Section B

	x_i	$x_i - \overline{x}$	$(x_i - \overline{x})^2$
	50	$50 - 75 = -25$	$(-25)^2 = 625$
	50	$50 - 75 = -25$	$(-25)^2 = 625$
	50	$50 - 75 = -25$	$(-25)^2 = 625$
	75	$75 - 75 = 0$	$(0)^2 = 0$
	100	$100 - 75 = 25$	$(25)^2 = 625$
	100	$100 - 75 = 25$	$(25)^2 = 625$
	100	$100 - 75 = 25$	$(25)^2 = 625$
$\Sigma(x_i - \overline{x})^2$			3750

So,

$$\sigma = \sqrt{\frac{\Sigma(x_i - \overline{x})^2}{n}} = \sqrt{\frac{3750}{7}} \approx 23.1$$

for Section B. This means that the average distance of the scores from the mean, 75, is roughly 23.1 points. Since the standard deviation for Section B is greater than the standard deviation for Section A, we know that the test scores for Section B are more spread out than those for Section A. ∎

Another measure of the spread of a data set uses percentiles or quartiles. The **nth percentile** of a distribution is the number such that n percent of the observations fall at or below it. Percentiles are frequently used to describe standardized test scores such as the ACT or the SAT. The median is always the 50th percentile. Other commonly reported percentiles are the 25th and the 75th. These percentiles are also called quartiles. The 25th percentile is known as the first quartile and the 75th percentile is known as the third quartile. (The median is the 50th percentile or the second quartile.) Just as the median divides a distribution into two groups with the same number of data points in each group, the 1st, 2nd, and 3rd quartiles divide a distribution into four groups with the same number of data points in each group. The 1st quartile can be thought of as the "middle" of the values below the median while the 3rd quartile can be thought of as the "middle" of the values above the median.

The difference between the 3rd quartile and the 1st quartile is called the **interquartile range.** This is also a measure of the spread of the data. For the test scores from Section A used in Example 2, the first quartile is 70, the second quartile (or median) is 75, and the 3rd quartile is 80.

50   **70**   70   **75**   80   **80**   100

1st quartile   median   3rd quartile

The interquartile range is $80 - 70 = 10$ for Section A.

For the test scores Section B, the first quartile is 50, the second quartile (or median) is 75, and the 3rd quartile is 100.

50	**50**	50	**75**	100	**100**	100
	1st quartile		median		3rd quartile	

The interquartile range is $100 - 50 = 50$ for Section B. The interquartile range shows that the scores for section B are more spread out than those for section A.

Normal Distributions

Figures 5.5, 5.6, and 5.7 are histograms for different data sets. For each histogram, we also computed the mean ($\overline{x}$) and the standard deviation (σ), then marked on the x-axis of each histogram where the mean ($\overline{x}$) occurs, and where the six points $\overline{x} \pm \sigma$, $\overline{x} \pm 2\sigma$, and $\overline{x} \pm 3\sigma$ lie. Figure 5.5 shows the chest sizes (in inches) of 5738 Scottish militiamen in the early 19th century.

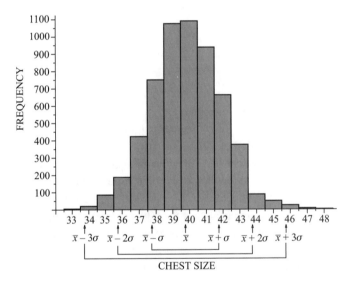

FIGURE 5.5
Histogram showing relative frequency for the chest sizes of Scottish Militiamen in the early 19th century. The mean is 39.8, the standard deviation is 2.0, and the total number of observations is 5738.

Figure 5.6 shows ages of 60 CEO's of small companies. Figure 5.7 shows the breaking strength of 100 samples of yarn. The breaking strength in pounds measures how many pounds it takes to break the yarn sample.[4]

Looking at each of the histograms, we can see that they are similar in shape even though they measure quite different things. Most of the data is clustered in the middle with small bars at either extreme, resulting in a "bell" shape. Each of the histograms can

---

[4] *Distribution Patterns*, 19 June 2002, <http://lib.stat.cmu.edu/DASL/Datafiles/distributiondat.html>.

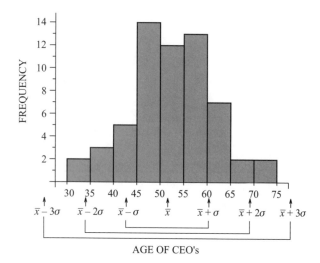

FIGURE 5.6

Histogram showing the age of CEO's of small companies. The mean is 51.7, the standard deviation is 8.8, and total number of observations is 60.

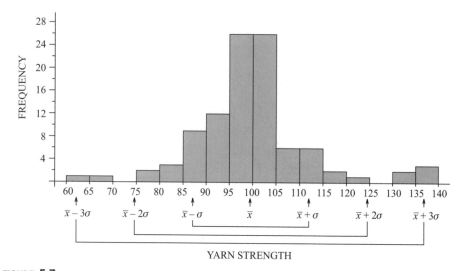

FIGURE 5.7

The breaking strength of yarn given in pounds. The mean is 99.43, the standard deviation is 12.46, and total number of observations is 100.

be approximated with the normal curve shown in Figure 5.8. The **normal curve** is a mathematical model for a normal distribution. We use the normal curve as an idealized model for describing these types of histograms. Histograms that are bell shaped can be described with just two numbers: the mean (measure of center) and the standard deviation (measure of spread). While we defined mean and standard deviation for a set of n values, mathematically there is a way to compute (using calculus) the mean and standard deviation

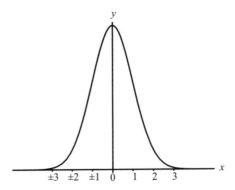

FIGURE **5.8**
A graph of the normal curve. This is a mathematical model for a normal distribution.

for continuous functions. For the normal distribution shown in Figure 5.9, the mean ($\bar{x}$) is 0 and the standard deviation (σ) is 1. The lines in Figure 5.9 represent the six points $\bar{x} \pm \sigma$, $\bar{x} \pm 2\sigma$, and $\bar{x} \pm 3\sigma$. Since the mean is 0 and the standard deviation is 1, these points occur at 0, ±1, ±2, ±3.

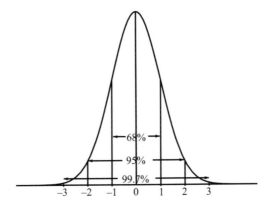

FIGURE **5.9**
The normal curve has the properties that 68% of the data are within one standard deviation of the mean, 95% are within two standard deviations of the mean, and 99.7% are within three standard deviations of the mean.

Useful properties of the normal distribution are:

- About 68% of the observations lie within one standard deviation of the mean,
- About 95% of the observations lie within two standard deviations of the mean,
- About 99.7% of the observations lie within three standard deviations of the mean, and
- Observations more than three standard deviations from the mean are exceedingly rare.

These properties are also true for data whose shape is similar to a bell-shaped curve. For example, the histograms we looked at earlier (chest size, age of CEO's, and yarn strength) follow these rules. So about 68% of the data are within one standard deviation of the mean, about 95% are within two standard deviations, and all (or almost all) of the data are within three standard deviations. Let's illustrate this using the militia example shown in Figure 5.5. The table of actual values is given below. There are 5738 values in all.

Chest Size (inches)	33	34	35	36	37	38	39	40	41	42	43	44	45	46	47	48
Frequency	3	18	81	185	420	749	1073	1079	934	658	370	92	50	21	4	1

The mean is 39.8 and the standard deviation is 2.0. If we count the values within one standard deviation (2.0) of the mean (39.8), so from 37.8 to 41.8, we get $749 + 1073 + 1079 + 934 = 3835$. This is $\frac{3835}{5738} = 66.8\%$ of the values (close to 68%). Similarly, within two standard deviations of the mean is from 35.8 to 43.8. Counting these values gives us $185 + 420 + 749 + 1073 + 1079 + 934 + 658 + 370 = 5468$ which is $\frac{5468}{5738} = 95.3\%$ of the values. Within three standard deviations of the mean is from 33.8 to 45.8. This corresponds to $18 + 81 + 185 + 420 + 749 + 1073 + 1079 + 934 + 658 + 370 + 92 + 50 = 5709$ which is $\frac{5709}{5738} = 99.5\%$ of the values. These numbers (66.8% within one standard deviation, 95.3% within two standard deviations, and 99.5% within three standard deviations) correspond quite well with the behavior of the normal curve.

Application: Calorie Content

Breakfast is said to be the most important meal of the day, but do you know how your breakfast cereal compares to others? The histogram in Figure 5.10 represents the calorie content of one serving of 50 popular breakfast cereals.[5]

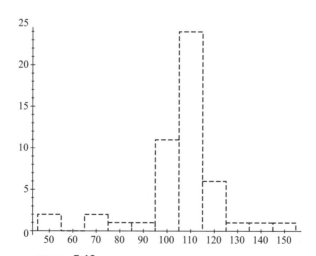

FIGURE 5.10
Calorie distribution for 50 popular breakfast cereals.

---

[5] *Cereals Datafile*, 20 June 2002, <http://lib.stat.cmu.edu/DASL/Datafiles/Cereals.html>.

We can estimate the mean and median content by looking at the graph. The graph is fairly symmetric (perhaps a bit skewed to the left) and centered around 110. We can accurately guess that the median calorie content will most likely be 110, since about half of the cereals surveyed contained 110 calories per serving and there are a few below this and a few above this number. Will the mean be 110 as well? To estimate the mean, we will have to look at all the data points, especially the outliers, since we know outliers have an effect on the mean. The graph seems to be slightly skewed to the left, so we can assume that the mean will be pulled slightly to the left. This suggests the mean will be slightly less than the median.

Example 3. The following table is a frequency distribution for the original 50 cereals. Using this distribution, calculate the mean and median calorie counts for the cereals.

Calories	50	60	70	80	90	100	110	120	130	140	150
Frequency	2	0	2	1	1	11	24	6	1	1	1

Solution. Since this is an even numbered data set, we take the average of the two middle numbers (the 25th and 26th data point) to find the median calorie content. By looking at the distribution, we see these two points are the same, 110, so their mean will be the same as well. Therefore, the median calorie content is 110, exactly what we predicted it would be. To find the mean, we add up the number of calories in each cereal and divide by the total number of cereals. To shorten this calculation, we note from our histogram that several cereals share the same calorie content. So we can multiply the total number of cereals in each category by the number of calories for that category. Therefore, the mean is

$$\bar{x} = \frac{(50\times2) + (70\times2) + 80 + 90 + (100\times11) + (110\times24) + (120\times6) + 130 + 140 + 150}{50}$$
$$= 105.8.$$

Our mean calorie content is slightly smaller than the median. As expected, the extreme data points on the left side of the graph pulled the mean to the left. ■

This shows the importance of knowing and recognizing the distinction between mean and median. While they both measure center, they have different interpretations of how the center of the distribution is defined.

Reading Questions for Describing Data

1. Find the median for each of the following.

   (a) $33, 42, 16, 25, 29, 30$

   (b) $98, 83, 76, 58, 95, 87, 91$

   (c) What did you do differently for part (b) as compared to part (a)?

2. The number of home runs hit by Babe Ruth each season (1920 to 1934) was:[6]

$$54, 59, 35, 41, 46, 25, 47, 60, 54, 46, 49, 46, 41, 34, 22$$

Find the mean and median.

3. Compute the mean and median for each of the following sets of numbers.

   (a) $3, 21, 11, 14, 5, 8, 5$

   (b) $-10, 4, 15, -3, 8, 1$

4. A fund-raiser for an elementary school organization brings in the following amounts of money:

$$\$60, \$125, \$80, \$25, \$130, \$90, \$50, \$75, \$40, \$140$$

All the money is to be used for a trip and the adult leader decides that each of the 10 children will get the same amount of money. How much does each child receive?

5. List ten numbers that have exactly one outlier.

6. List five numbers whose mean is smaller than the median.

7. List ten numbers whose histogram will be skewed to the right. Sketch the histogram for your data.

8. Suppose there are nine people working in a dentist's office. The three people that keep track of the patients and the books each earn $36,000 per year, the four dental hygienists each earn $42,000 per year, and the two dentists each earn $106,000 per year. What is the mean and the median salary of people working in this office? Which is the better measure of center in this case? Explain.

9. The following table contains the names and heights of the 2001–2002 University of Connecticut women's varsity basketball team.[7] Use it to answer the following questions.

Player	Height
Ashley Valley	5'9"
Diana Taurasi	6'0"
Maria Conlon	5'9"
Sue Bird	5'9"
Stacey Marron	5'9"
Asjha Jones	6'2"
Morgan Valley	6'0"
Ashley Battle	6'0"
Jessica Moore	6'3"
Swin Cash	6'2"
Tamika Williams	6'2"

---

[6] *The Official Babe Ruth Website: Statistics*, 19 June 2002, <http://www.baberuth.com/stats.html>.

[7] *www.UConnHuskies.com Women's Basketball*, 19 June 2002, <http://www.uconnhuskies.com/sports/WBasketball/2002/Roster/roster2001-2002.html>.

(a) Without doing any calculations, which of the following is most likely to be true? Briefly justify your answer.

- The mean will be greater than the median.
- The mean will be less than the median.
- The mean will be almost equal to the median.

(b) Compute the mean and the median for this data.

10. When is it better to use the median instead of the mean as a measure of center?

11. Points A and B are labeled on the following histograms. Determine if A represents the mean, median, both, or neither. Do the same for B.

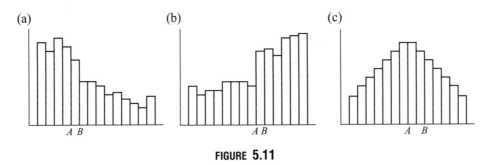

FIGURE 5.11

12. If the variance for a data set is 441, what is the standard deviation?

13. The prices of ten used cars are given below.[8]

Year	Model	Price
1993	Mercury Sable	$2595
1994	Plymouth Voyager	$4980
1995	Dodge Ram 2500	$8995
1995	GMC Yukon	$10,500
1996	Dodge Neon	$5998
1996	Dodge Caravan	$6888
1997	Ford Mustang	$16,995
1998	Chrysler Concorde	$10,995
1999	Dodge Intrepid	$11,995
1999	Ford Explorer	$13,995

Compute the mean, median, and the standard deviation using a calculator or software. Describe the meaning of each of these numbers.

14. Compute the range and the standard deviation for the following set of numbers.

$$11, 9, 12, 4, 2, 8$$

---

[8] *Yahoo Auto Classifieds*, 19 June 2002, <http://classifieds.autos.yahoo.com/display/automobiles?ct_hft=browse &action=browse& nodeid=750000911&srid=375000157&intl=us>.

15. What does it mean to be in the 85th percentile for ACT scores?

16. If a data set has a large interquartile range, what does this say about the spread of the data?

17. For the following data set, find $minX$ (minimum value), Q_1 (1st quartile), Med (median), Q_3 (3rd quartile), and $maxX$ (maximum value).

$$168, 152, 210, 177, 158, 180, 193, 165$$

18. Find the standard deviation using a calculator or software and the interquartile range for the following sets of data:

    (a) $20, 20, 21, 23, 26, 27, 27, 30, 31, 31, 32$

    (b) $10.1 \quad 10.8 \quad 12.6 \quad 13.1 \quad 14.3 \quad 14.4 \quad 14.4 \quad 14.6 \quad 14.7 \quad 15.1 \quad 16.0 \quad 16.3 \quad 16.8 \quad 18.1$

19. Using your calculator or software, find the one-variable statistics of the following frequency distribution of student heights. Write a brief sentence describing each of the results.

Height	66	67	68	69	70	71	72
Frequency	1	3	5	8	7	3	2

20. The mean IQ is 100 and the standard deviation is 15.[9] IQ follows a normal distribution. What is the range of IQ scores for the middle 95% of the population?

21. The following table gives the breaking strength of 100 samples of yarn. (This was used to produce the histogram in the reading.) The mean is 99.43 and the standard deviation is 12.46. Show that this example follows the 68% - 95% - 99.7% rule.

Breaking Strength	Frequency (Number of Observations)	Breaking Strength	Frequency (Number of Observations)	Breaking Strength	Frequency (Number of Observations)
62	1	93	1	106	1
66	1	94	3	107	2
78	1	95	3	109	1
79	1	96	5	110	1
80	1	97	6	111	4
84	2	98	7	114	1
85	2	99	5	115	1
86	2	100	5	117	1
87	1	101	4	122	1
88	2	102	7	132	2
89	2	103	3	137	2
91	4	104	7	138	1
92	4	105	2		

---

[9] *The Intelligence Quotient*, 12 March 2004, <http://www.psyonline.nl/en-iq.htm>.

22. Suppose the calorie content of two new cereals are added to the histogram used in Example 3 and shown below. One is "Healthy Puffed Corn" whose calorie count for one serving is 50. The other is "Crunchy Fudge Roasted Pecan Clusters" whose calorie count for one serving is 300. How will this affect the mean and median calorie counts for the cereals? Explain.

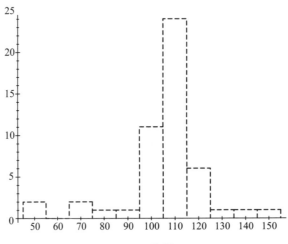

FIGURE 5.12

Describing Data: Activities and Class Exercises

1. **100 Best Thrillers.** In 2001, the *American Film Institute* released a list of what it called the 100 best thrillers.[10] Below is a graph depicting the years that the films were released.

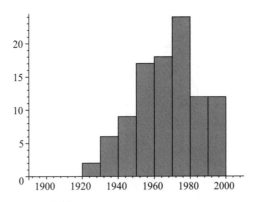

FIGURE 5.13
American Film Institute's 100 Years, 100 Thrills

---

[10] *American Film Institute 100 Years, 100 Movies*, 25 June 2002, <http://www.afi.com/tv/thrills.asp>.

(a) What type of graph is this?

(b) According to the *American Film Institute*, which decade had the most "best thrillers"?

(c) How many "best thrillers" were released between 1980 and 2000?

(d) List five movies from 1980 to 2000 that you think were considered "best thrillers" by the *American Film Institute*. [Note: Remember that this list was released in June 2001.]

(e) Estimate the median year for the 100 best thrillers. Describe what your median represents in terms of the 100 best thrillers.

(f) Suppose each decade had the same number of movies on the list. In this case, what would be the shape of the histogram? Why do you think the number of films is not the same for each decade?

2. **Astros' Salaries.** The following table shows the salary for members of the Houston Astros baseball team on opening day in 2003.[11]

Player	Salary	Player	Salary
Ausmus, Brad	$5,500,000	Lidge, Brad	$300,000
Bagwell, Jeff	$13,000,000	Miceli, Danny	$350,000
Berkman, Lance	$3,500,000	Miller, Wade	$525,000
Biggio, Craig	$9,750,000	Oswalt, Roy	$500,000
Chavez, Raul	$300,000	Palmeiro, Orlando	$700,000
Clemens, Roger	$7,061,181	Pettitte, Andy	$11,500,000
Dotel, Octavio	$1,600,000	Redding, Tim	$300,000
Duckworth, Brandon	$325,000	Robertson, Jeriome	$300,000
Ensberg, Morgan	$300,000	Stone, Ricky	$325,000
Fiore, Tony	$330,000	Veres, Dave	$2,000,000
Hidalgo, Richard	$8,500,000	Vizcaino, Jose	$2,000,000
Kent, Jeff	$6,949,840		

(a) Without doing any calculations, decide if the median salary or the mean salary will be higher. Explain how you determined your answer.

(b) Calculate the mean, median, and standard deviation of the salaries for the 2003 Houston Astros using a calculator or software.

(c) Assume players on the Astros are paid according to their abilities. What should a player of average ability expect to get paid? Explain your answer.

(d) Suppose each player on the Houston Astros was given a $50,000 pay raise.

---

[11] *MLB-Houston Astros Roster-CBS.SportsLine.com*, 12 March 2004, <http://www.sportsline.com/mlb/teams/roster/HOU.>

Without recalculating, determine the new mean, median, and standard deviation. Explain how you determined your answers.

3. **Growth Chart.** The National Center for Health Statistics Growth Chart is used by doctors to chart the length (i.e., height) and weight of children from birth to 36 months. The weights are given in pounds and kilograms and the lengths are given in inches and centimeters. Use the chart in Figure 5.14 to answer the following questions.

   (a) In what percentile is a 12-month old child who is 24 pounds?

   (b) In what percentile is a 3-month old child who is 24 inches long?

   (c) Suppose you have a 9-month old child who is 17.5 pounds and 30 inches tall. How would you describe the weight and height of your child when compared to other children his or her age?

   (d) If a 24-month old child is at the 50th percentile in both height and weight, what is his or her height and weight?

   (e) The 50th percentile for the height of a 24-month old child is about 34.5 inches. What is the median height for a child this age?

   (f) During what age interval is the rate of weight gain the largest? What is the rate in pounds per month?

   (g) As children get older, their average weight increases. Does the standard deviation of weights also increase? How can you tell by looking at the chart?

4. **Standard Deviation.** Given the set of numbers $\{1, 2, 3, 4, 5\}$, you are to choose four of them with repetitions allowed.

   (a) Give a list of four of these numbers so that the standard deviation is as small as possible. Is your list unique or is there another list of four that will have the same standard deviation? Explain.

   (b) Give a list of four of these numbers so that the standard deviation is as large as possible. Is your list unique or is there another list of four that will have the same standard deviation? Explain.

5. **Exam Grades.** A college professor states that he will curve exams 10 percentage points if the class average is below 70. Suppose the test scores are:

   $$75, 51, 60, 97, 90, 30, 62, 89, 83, 78, 25, 77,$$
   $$80, 70, 82, 59, 83, 85, 61, 92, 78, 82, 90, 77, 78$$

   (a) What is the median test score?

   (b) What is the mean test score? Will the professor curve the exam?

   (c) What are the percentages for the first and third quartiles? What is the interquartile range?

   (d) What is the standard deviation? What does this tell you about the spread?

   (e) Are there any outliers? If so, list them.

   (f) Do you think the professor should change his syllabus and use the median test score instead of the mean for curving exams? Explain.

Birth to 36 months: Boys
Length-for-age and Weight-for-age percentiles

NAME _____

RECORD # _____

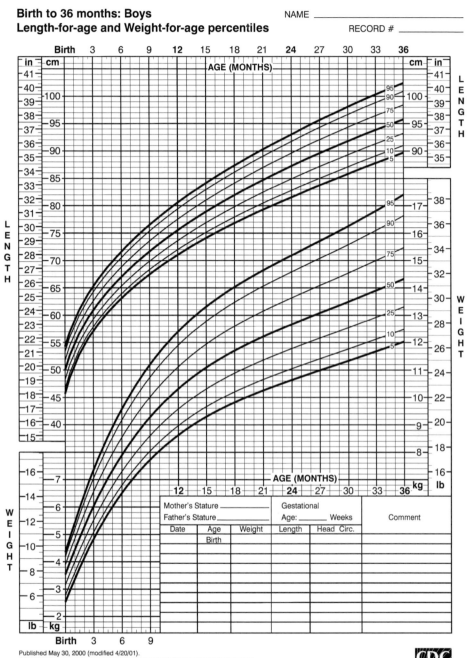

Published May 30, 2000 (modified 4/20/01).
SOURCE: Developed by the National Center for Health Statistics in collaboration with
the National Center for Chronic Disease Prevention and Health Promotion (2000).
http://www.cdc.gov/growthcharts

SAFER · HEALTHIER · PEOPLE™

FIGURE **5.14**

6

Multivariable Functions and Contour Diagrams

In earlier sections, we looked at functions and their graphs. However, we only looked at functions with one input and one output. But sometimes a function has more than one input. For example, consider the process of computing a monthly payment for a car loan. The payment depends on the price of the car, the interest rate, and the length of the loan. There are actually three inputs or independent variables (price, interest rate, and length of the loan) that determine the output (monthly car payment). In this section, we will look at functions with more than one input and their graphs.

Functions with Two Inputs

There are many examples of functions that have more than one input. One example of a function with two inputs is the formula for the volume of a cone, $V = \frac{1}{3}\pi r^2 h$, where h is the height of the cone, and r is the radius of the circular base. (See Figure 6.1.) This is an example of a multivariable function. A **multivariable function** is a function that has more than one input. In this text, we will only consider functions with two inputs and a single output so that one variable depends on two distinct independent variables. Notice that the volume formula is still a *function*. That is, for any pair of inputs (i.e., for any fixed height and radius), there is only one possible output (i.e., volume).

Example 1. Suppose a geologist wants to find a rough estimate of the volume of Mount St. Helens before its eruption on May 18, 1980. The shape of Mount St. Helens was roughly a cone with a height, h, of 1.83 miles and a diameter at the base of 6 miles.[1]

---

[1] *Mount St. Helens—From the 1980 Eruption to 2000, "Fact Sheet,"* 24 March 2004, <http://geopubs.wr.usgs.gov/fact-sheet/fs036-00/>.

FIGURE 6.1
The volume of a cone is $V = \frac{1}{3}\pi r^2 h$, where h is the height and r is the radius of the circular base.

Use the formula for the volume of a cone to estimate the volume of Mount St. Helens before its eruption.

Solution.   One of the inputs, height (h), was 1.83 miles. We now have to find the other input, the radius of the base of the volcano (r). Since the radius is half the diameter, we use half of 6 or 3 miles as the radius. We can now estimate the volume of Mount St. Helens by using the formula for the volume of a cone.

$$V = \frac{1}{3}\pi r^2 h$$

$$V \approx \frac{1}{3} \times 3.14 \times 3^2 \times 1.83$$

$$V \approx 17.24 \text{ cubic miles.}$$

■

   Another example of a multivariable function is the formula used to compute the monthly payment for a loan. Suppose you are planning on taking out a five-year (or 60-month) loan to buy a car. To determine your monthly payment, the price of the car is important, but so is the interest rate. Both of these are used as variables in the standard formula that computes the monthly payment, M, for a 60-month loan:

$$M = \frac{iL}{1-(1+i)^{-60}}, \tag{1}$$

where i is the monthly interest rate and L is the amount of the loan.[2] There are 2 inputs (i and L) and one output (monthly payment).

Example 2.   Suppose we want to take out a five-year loan for \$16,000 with a yearly interest rate of 9%. Determine the monthly payment.

Solution.   Before using Formula (1) to compute the monthly payment, we have to convert the yearly interest rate to a monthly interest rate by dividing the 9% by 12. Using

---

[2] The 60 in this equation comes from making 60 payments, one payment per month for 5 years.

$L = 16,000$ and $i = \frac{0.09}{12}$, we get

$$M = \frac{\frac{0.09}{12} \cdot 16000}{1 - (1 + \frac{0.09}{12})^{-60}} \approx \frac{120}{1 - 0.6387} \approx 332.13.$$

So our monthly car payment is $332.13. ∎

Let's look at the car payment example again, changing the notation slightly. Instead of writing

$$M = \frac{iL}{1 - (1 + i)^{-60}},$$

we will now write

$$M(i, L) = \frac{iL}{1 - (1 + i)^{-60}}.$$

This notation is similar to the notation we used in the *Functions* chapter where we saw that $y = x^2$ and $f(x) = x^2$ were two ways of representing the same function. In the case of the monthly payment function, the notation $M(i, L)$ reminds us that M, the monthly payment, is the output and that i, the interest rate, and L, the amount of the loan, are the inputs.

Example 3. Consider the formula for computing the monthly payment on a 60-month loan,

$$M(i, L) = \frac{iL}{1 - (1 + i)^{-60}}.$$

Give the formula for $M(\frac{0.09}{12}, L)$ and explain what this new function represents.

Solution. $M(\frac{0.09}{12}, L)$ means substitute $\frac{0.09}{12}$ in place of i in the formula for $M(i, L)$. Doing this gives us

$$M\left(\frac{0.09}{12}, L\right) = \frac{\frac{0.09}{12} L}{1 - (1 + \frac{0.09}{12})^{-60}}.$$

This represents the monthly payment for a 60-month loan of amount L assuming a monthly interest rate of $\frac{0.09}{12}$ (or a 9% annual rate). ∎

Example 3 demonstrates how to create a one-variable function from a two-variable function by fixing one of the inputs as a constant. Instead of letting both inputs vary, we substitute a number for one of the inputs. In Example 3, the interest rate was fixed and the amount of the loan varied. We could also hold the amount of the loan constant and let the interest rate vary to create a different one-variable function.

Let's consider another two-variable function. Table 6.1 represents the two-variable function for computing the wind-chill equivalent temperature.[3] This function has two inputs (wind speed and air temperature) and one output (wind-chill). Let $W(s, t)$ represent this function where W is the wind-chill equivalent temperature in degrees Fahrenheit, s is the wind speed in mph, and t is the temperature in degrees Fahrenheit.

---

[3] *NWS Wind Chill Temperature Index*, 24 March 2004, <http://www.nws.noaa.gov/om/windchill/>.

TABLE 6.1

Wind-Chill Equivalent Temperature

Wind Speed (mph)	Air Temperature (°F)												
	35	**30**	**25**	**20**	**15**	**10**	**5**	**0**	**−5**	**−10**	**−15**	**−20**	**−25**
5	31	25	19	13	7	1	−5	−11	−16	−22	−28	−34	−40
10	27	21	15	9	3	−4	−10	−16	−22	−28	−35	−41	−47
15	25	19	13	6	0	−7	−13	−19	−26	−32	−39	−45	−51
20	24	17	11	4	−2	−9	−15	−22	−29	−35	−42	−48	−55
25	23	16	9	3	−4	−11	−17	−24	−31	−37	−44	−51	−58
30	22	15	8	1	−5	−12	−19	−26	−33	−39	−46	−53	−60
35	21	14	7	0	−7	−14	−21	−27	−34	−41	−48	−55	−62
40	20	13	6	−1	−8	−15	−22	−29	−36	−43	−50	−57	−64

Example 4.   Compute $W(20, 10)$.

Solution.   Use the left column of Table 6.1 to find $s = 20$ for the wind speed and use the top row to find $t = 10°$ for the air temperature. At their intersection, we see that the wind-chill equivalent temperature is $−9°$. (Rather chilly!!) ■

Caution: When evaluating a multivariable function, be consistent in the notation. If the function is given as $W(s, t)$, then the number substituted for the first input, s, must *always* be wind speed while the number substituted for the second input, t, must *always* be air temperature.

Example 5.   Explain the meaning of $W(s, 5)$.

Solution. $W(s, 5)$ is the one-variable function whose input is wind speed and whose output is the wind-chill equivalent temperature when the air temperature is fixed at 5° Fahrenheit. ■

Interpreting Contour Maps

Topographical Maps

We see graphs of functions of two variables in a variety of places. Two common examples are topographical maps (commonly used by geologists and hikers) and weather maps (commonly given in newspapers). In a typical topographical map, we define $z = E(x, y)$ as the function whose input is the latitude, x, and longitude, y, of the position and whose output, $E(x, y)$, is the elevation or height of that position above sea-level. This function, $z = E(x, y)$, is a two-variable function since there are two inputs, x and y. Figure 6.2 is a diagram of a hill and some of the heights associated with it.

The first picture shows the hill as if you were looking up at it from the ground. This is the perspective that we are most accustomed to seeing. The picture immediately below

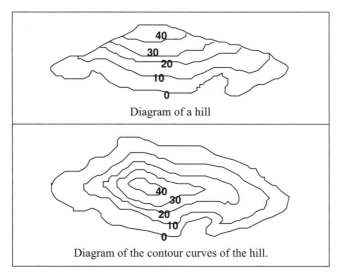

Diagram of a hill

Diagram of the contour curves of the hill.

FIGURE 6.2
Diagram of a hill and its contour curves.

it shows the hill as if we were looking straight down at it. The labelled curves on the two pictures are the same. Each curve, called a **contour**, represents all points that are at a given height above some reference level. The numbers on the contour curves describe the height (output) for the function $z = E(x, y)$. For example, the contour curve for 40 is much shorter than the contour curve for 10. This is because 40 is near the top of the mountain and so there is a relatively small cross-section of the mountain exactly 40 units above sea level. On the other hand, 10 is close to the bottom of the mountain and so there is a relatively large cross-section of the mountain exactly 10 units above sea level.

The contour maps for terrains consisting of a hill and a cliff are given in Figure 6.3. The hill has a gradual ascent while the cliff is quite steep and changes elevation very quickly. Can you tell which one is the hill and which one is the cliff?

The cliff is on the left (Figure 6.3(a)) because of the cluster of contour curves. When the contour curves are close together, a small change in latitude and longitude is causing a rapid change in elevation. Hence, the closer together the contour curves, the steeper the

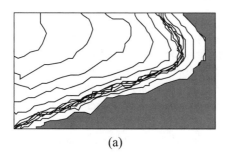

(a)

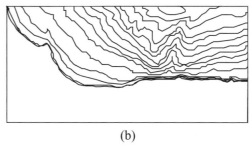

(b)

FIGURE 6.3
One of these contour maps is of a cliff and one is of a hill. Which is which?

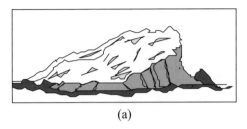
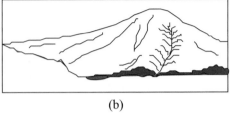

(a)   (b)

FIGURE 6.4
The cliff in (a) is the three-dimensional visualization of the contours in Figure 6.3(a). The hill in (b) created the contours in Figure 6.3(b).

terrain. The top of the cliff is fairly flat since the contour curves on the left-side of Figure 6.3(a) are spread out. The hill is the figure on the right (Figure 6.3(b)). Three-dimensional views for each of these terrains are shown in Figure 6.4.

Weather Maps

Looking at contour curves on a topographical map is similar to looking at the temperature curves drawn on a weather map.[4] Let $z = T(x, y)$ represent the function whose input is latitude, x, and longitude, y, and whose output, $T(x, y)$, is the current (or high) temperature for that day at the point (x, y). The curves on the weather map separating the temperature regions indicate all the places where the temperature (output) has the same fixed value. These curves are actually contour curves for $z = T(x, y)$.

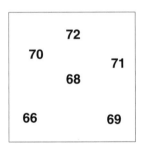

FIGURE 6.5
A small portion of a weather map showing temperatures for the area.

Figure 6.5 represents a small portion of a weather map showing local temperatures. Generally, weather maps have contour curves drawn at locations rounded to the nearest ten degrees (i.e., 50°, 60°, 70°). We are interested in drawing the contour curve for 70°. This is the contour that, theoretically, "connects" all the locations where the temperature is exactly 70°. As you can see in Figure 6.5, our map contains selected temperatures. So not all of the points where the temperature is exactly 70° are marked. Instead, we have a few temperatures that are more than 70°, a few less that 70°, and only one that is exactly 70°.

---

[4] These curves on a weather map are called *isotherms*, meaning equal temperatures.

(a) 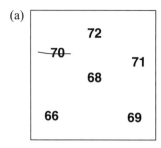 (b)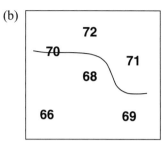

FIGURE 6.6
The beginning of a contour curve at 70° is drawn in (a). In (b), the entire 70° contour curve is drawn.

We will begin to draw our contour curve by starting with points that are exactly 70°. We know our curve must go through these points as shown in Figure 6.6(a). The next question is where this line should continue. To determine this, we must look at the surrounding temperatures in the region. As you can see, none of the surrounding temperatures are exactly 70°. However, there are a few temperatures close to 70°. We know that temperatures do not just "jump" between temperatures that are a few degrees apart. Rather, they progress continuously between the two. This means that between two locations that are 68° and 72°, there must be a location where the temperature is 70°. So we draw our contour curve between temperatures that are slightly higher than 70° and those that are slightly lower than 70° as shown in Figure 6.6(b).

Figure 6.7 represents an inaccurate 70° contour curve. The contour for 70° must go between temperatures of low 70s and high 60s. We know that 70° must lie between 69° and 71° because temperatures do not "jump" from one value to another. Figure 6.7 has the contour going between 71° and 72°, which is incorrect.[5]

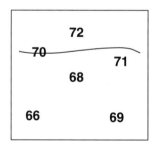

FIGURE 6.7
An incorrect contour curve. The contour for 70° must go between temperatures in the low 70s and high 60s.

---

[5] It is true that there could be a 70° temperature between a location with a temperature of 71° and another of 72° if the temperature was decreasing between these two readings. However, we are taking a more simplistic approach for our examples by assuming our readings take into account this type of behavior. Therefore, we will assume the temperature between any two points on a map will be between the two given temperatures.

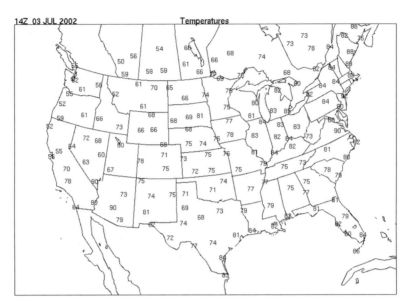

FIGURE 6.8
A weather map showing only temperatures for July 3, 2002.

A weather map for the United States is shown in Figure 6.8. Contour maps drawn by a weather service actually use many more temperatures than are displayed on the map and then "smooth" the contour curves. Unfortunately, this means that, on official maps, there will often be points that should be included on a contour (or isotherm) that are not.

In this book, we will assume that the only temperatures are the ones given on the map and we will estimate the contours accordingly. Figure 6.9 is the same weather map as that shown in Figure 6.8, this time with isotherms. Compare it to Figure 6.8 to see approximately how the isotherms, or contours, were determined.[6]

Application: Weather Map

We have seen that contours, called isotherms, are used in meteorology to show temperatures in a region. Figure 6.10 shows a United States weather map, taken from the *New York Times*, for January 13, 1998.

The interiors of these contours are shaded to make it easier for the reader to see temperatures for the areas they are interested in. For example, Chicago seems to be right on the line between 10s and 20s. So its temperature must be in the high 10s or the low 20s. Since it lies on the contour between the two, we will estimate its temperature to be about 20°. Cincinnati is near the contour separating the 20° temperatures and the 30° temperatures, so its temperature must be in the high 20s or low 30s. It lies more on the edge of the 30s, so we will estimate its temperature to be about 31°.

---

6 *AMS The DataStreme Project*, 1 July 2002, <http://www.ametsoc.org/dstreme/images/sfc_temp.gif>.

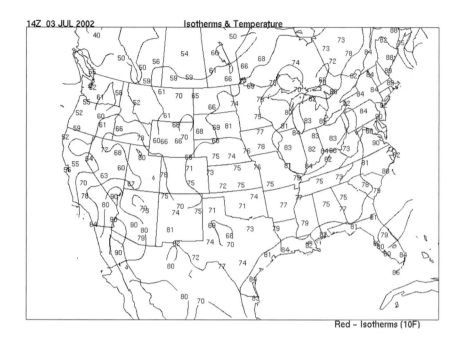

FIGURE 6.9
A weather map showing isotherms (contours) for July 3, 2002.

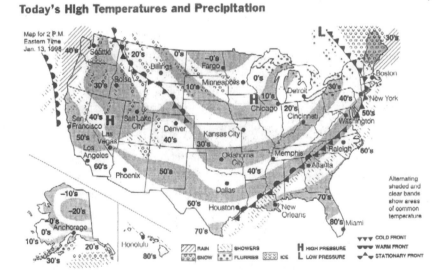

FIGURE 6.10
Weather map showing high temperatures for the United States on January 13, 1998.

Today's High Temperatures and Precipitation

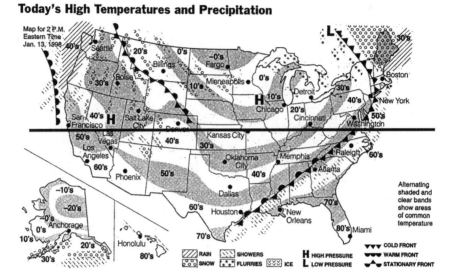

FIGURE **6.11**

Example 6.   An approximate east-west line is drawn through Kansas City on the map in Figure 6.11. Sketch a graph of the one-variable function whose input is miles west or east of Kansas City and whose output is temperature. Your domain should be from the west coast to the east coast with Kansas City located at the origin.

Solution.   Looking at the weather map, we can see that Kansas City is roughly between the 20° contour and the 30° contour. We will estimate its temperature to be about 25°. Now we need to look at the temperature regions that the line goes through, starting in the west.

In California, it starts out in the mid 50s, then cools down into the 40s in Nevada, and into the 30s in eastern Nevada and western Utah. But then the temperature increases back into the 40s again in the eastern half of Utah and in Colorado. The temperature turns colder again into the 30s for a short period in extreme western Kansas, and then it becomes even colder, in the 20s, through most of Kansas. Remember, we estimated Kansas City's temperature to be about 25°, so Kansas City should be our low point. The temperature should start to rise through the upper twenties in Missouri, Illinois, and Indiana, before hitting 30 in northern Kentucky, close to Cincinnati where earlier we predicted the temperature to be about 31°. The temperatures continue to rise in the 40s in West Virginia, and finally back into the mid to high 50s in Virginia on the east coast. A final graph of this function is shown in Figure 6.12.

This bowl-shaped trend in east-west temperatures is caused by the jet stream. The temperature graph reflects a cold front dipping down from Canada, making the interior portions of the United States much cooler than the coastal regions. ∎

Example 6 illustrates one way of making a multivariable function into a single variable function. The original temperature function, $z = T(x, y)$, depended on both the latitude and the longitude of our location. By drawing an east-west line through Kansas City, we

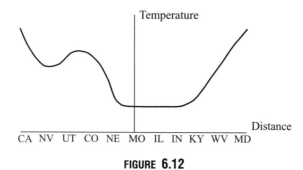

FIGURE 6.12

held the latitude constant, resulting in a single variable function. Changing a multivariable function into a single variable function by holding one of the inputs constant is what allowed us to make a two-dimensional graph of our answer (Figure 6.12).

Reading Questions for Multivariable Functions

1. What do we call a function with more than one input?

2. Give an example of a multivariable function different from those mentioned in the reading. Carefully define the input and the output variables.

3. Let $f(x, y) = x^2 + 2y$. Explain what $f(1, 2)$ means and then compute its value.

4. In Example 1, we estimated the volume of Mount St. Helens using the formula for a cone. In 1980, the eruption of Mount St. Helens blew the top 0.25 miles off the mountain. If we think of the portion of the mountain that was removed as a cone, it would have a base of approximately 0.82 miles and a height of 0.25 miles.

   (a) Use the function $V(r, h) = \frac{1}{3}\pi r^2 h$ to determine the volume of the portion of Mount St. Helens that was removed during the eruption.

   (b) While your answer in part (a), in cubic miles, may seem like a small number, a huge amount of earth was moved during the eruption. To put this in perspective, suppose we were to pile this amount of earth onto a football field with sides going straight up. An American football field is 100 yards long and 160 feet wide and has an area of approximately 0.00172 square miles. This means that $0.00172 \times$ [the height of the pile] = [your answer from part (a)]. Find the height of this pile.

5. Consider the function giving the volume of a cone, $V(r, h) = \frac{1}{3}\pi r^2 h$, and the function $z = V(3, h)$. Give the formula for $V(3, h)$ and explain what this new function represents.

6. Use the function
$$M(i, L) = \frac{iL}{1 - (1 + i)^{-60}}$$
to compute the monthly car payment on a 60-month loan of $18,000 at 10% annual interest rate.

7. Adapt the formula for a 60-month loan to one for a 48-month loan and use it find the car payment in Question 6. How much of a difference is there in the monthly payment?

8. Give the formula for $M(i, 2300)$ and explain what it means.

9. Suppose the annual interest rate is 9% and you can afford a monthly payment of $250 to $300 on a 60-month loan. What price of a car should you consider?

10. Find $W(10, -5)$ using Table 6.1.

11. Explain the meaning of $W(25, t)$.

12. Using the wind-chill table in the reading, determine how fast the wind would need to blow to make it feel like $-7°$ if the air temperature is $15°$.

13. Explain how to create a new one-variable function from a two-variable function.

14. A 10-meal plan at a midwestern liberal arts college costs $1,350 for the semester. There are also two different commuter passes for lunch and dinner costing $55 and $65 respectively for **10 meals**.

    (a) Write a multivariable function where the input is the number of commuter lunch passes bought and the number of commuter dinner passes bought and the output is the total cost.

    (b) Which would be the better deal, a 10 meal plan, or commuter passes totaling 10 meals a week? Assume you eat 5 lunches and 5 dinners a week and the semester is 15 weeks.

15. Give two examples where contour curves are commonly used.

16. The following table shows annual life insurance payments for different plans (in thousands of dollars) for a non-smoking male, ages 31 to 40.[7]

Age	Insurance Plan (in thousands of dollars)									
	50	100	150	200	250	300	350	400	450	500
31	115.50	155.00	207.50	260.00	272.50	317.00	361.50	406.00	450.50	495.00
32	117.00	157.00	210.50	264.00	277.50	323.00	368.50	414.00	459.50	505.00
33	120.00	162.00	218.50	274.00	287.50	335.00	382.50	430.00	477.50	525.00
34	122.50	166.00	224.00	282.00	297.50	347.00	396.50	446.00	495.50	545.00
35	126.50	172.00	233.00	294.00	310.00	362.00	414.00	466.00	518.00	570.00
36	130.50	179.00	243.50	308.00	325.00	380.00	435.00	490.00	545.00	600.00
37	135.50	187.00	255.50	324.00	340.00	398.00	456.00	514.00	572.00	630.00
38	141.50	196.00	269.00	342.00	360.00	422.00	484.00	546.00	608.00	670.00
39	149.00	208.00	287.00	366.00	385.00	452.00	519.00	586.00	653.00	720.00
40	158.00	223.00	309.50	N/A	417.50	491.00	564.50	638.00	711.50	785.00

---

[7] *Term Life Insurance Rate Company Low Rates on Term Life Insurance Quotes*, 18 Jun 2002, <http://www.TermLifeInsuranceRateCompany.com>.

(a) What are the input(s) for this function? What are the output(s)?

(b) Suppose the customer is a thirty-eight year old man. The man has decided to budget 250 to 275 dollars each year towards a life insurance policy. What policy should he consider?

(c) The man thought about getting a policy seven years earlier when he was thirty-one. Assuming he had the same budget, what policy should he have considered?

(d) In general, when should you buy life insurance?

17. Briefly describe what the terrain is like at each of three points on the following contour map.

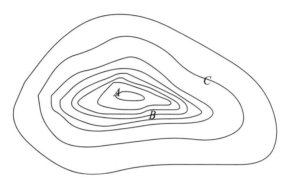

FIGURE **6.13**

18. Draw a possible contour curve (isotherm) for 70° for each the following regions.

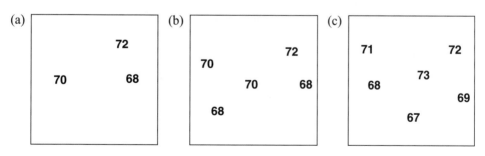

FIGURE **6.14**

19. Use the weather map in Figure 6.15 to answer the following questions.

(a) Estimate the high temperature in Chicago within 5 degrees.

(b) Trace the contour curve (isotherm) for $T = 60$ on the U.S. weather map.

20. Draw the isotherm for $T = 70°$ on the weather map in Figure 6.16.

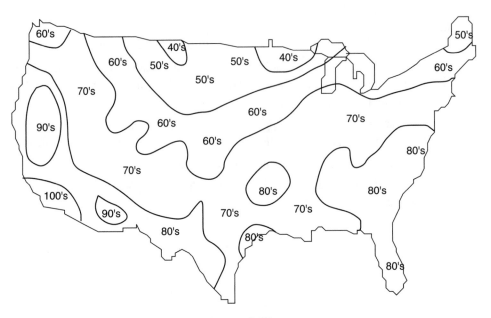

FIGURE **6.15**

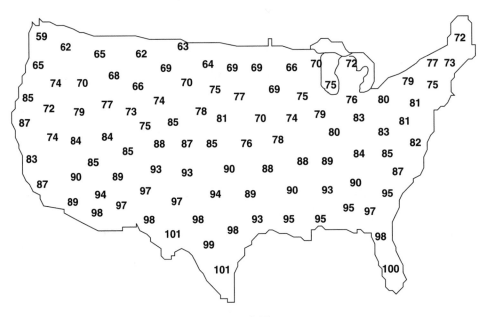

FIGURE **6.16**

Today's High Temperatures and Precipitation

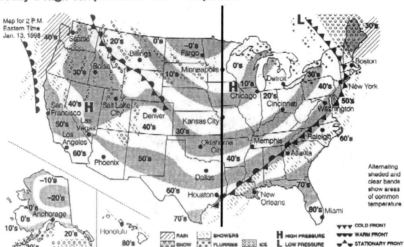

FIGURE **6.17**

21. Figure 6.17 is the map from Figure 6.10, with a north-south line included on it. Sketch a graph of the one-variable function whose input is miles north or south of Kansas City and whose output is temperature. Your domain should be from the Canadian border to the Gulf of Mexico with Kansas City located at the origin.

Multivariable Functions: Activities and Class Exercises

1. **Mount Rainier.** Use a topographical map of Mount Rainier to answer the following questions. The map can be acquired from Earthwalk Press, 2239 Union Street, Eureka, CA 95501 or ordered from <http://www.globecorner.com/s/400.html.>

   (a) Looking at the contours, describe the shape of the peak of Mount Rainier.

   (b) There are two trails leaving White River Campground, one going east-west and one going north-south. Which one will be the steeper climb? How do you know?

   (c) What is different about the contour curves on the glaciers versus the contour curves on the rest of the mountain? What does this tell you about the glaciers?

2. **Fish Populations.** Use the surface graph from D.K. Cairn's article on fish to answer the following questions.[8]

   (a) This is a graph of a multivariable function with two inputs and one output. Indicate which variables are inputs and which are outputs.

---

[8] Cairns D.K, "Bridging the Gap Between Ornithology and Fisheries," *The Condor*, 1992.

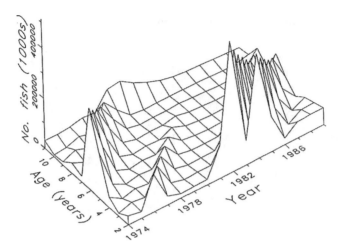

FIGURE **6.18**

(b) Which year saw the largest catch of fish for a single age of fish? What age were the fish in that catch?

(c) What age fish were most heavily caught in 1974?

(d) What year is an 11-year-old fish most likely to be caught?

3. **Modeling Clay.** Take four colors of modeling clay. Flatten these into rectangular layers of *equal thickness* and then stack them, trying not to mix colors. Use a table knife to cut the outside edges, creating something that resembles the shape of a cone or a mountain. (When you are cutting, it works better to use steady pressure, rather than a sawing motion. This will help you avoid smearing the contacts between layers.) Make sure that your sides have differing slopes, i.e., you'll want to make one side steeper than another. The contacts between the slabs of clay are your contour lines.

   (a) Looking at your "mountain" from above, sketch the contour lines, labeling the approximate elevations from the tabletop that each represents.

   (b) What can you determine about areas where the contour lines are closer together? What about where they are further apart?

   (c) Carve a "river" down a side of your mountain. Sketch your revised contour lines.

   (d) Can contour lines ever cross? Explain.

4. **Boyle's Law.** The mathematical equation that describes the relationship between the pressure and the volume for an ideal gas is known as Boyle's Law. It states that

$$PV = nRT$$

where P is pressure, V is volume, n is the amount of gas present, R is a constant associated with the gas, and T is the temperature.

(a) Rewrite Boyle's Law so that n, T, and V are your inputs while P is your output. [Note: R is constant so it is not considered to be an input.]

(b) Suppose we had the gas in a sealed container so that the volume, V, and the amount, n, stayed the same. If we increase the temperature, T, by heating the gas, what happens to the pressure? Does it increase or decrease? Explain.

(c) Suppose we wanted to heat the gas but also wanted to have the pressure stay the same. What would we have to do to the volume? Why?

(d) If you inflated a tire to its proper pressure on a very cold day, what would happen to the pressure if the weather suddenly turned warm? Explain.

5. **Car Loan.** Suppose you just graduated from college and decide to purchase a car. You have budgeted your living expenses and estimate you can afford $500 a month for car expenses. Assume the annual interest rate is 9.5%.

(a) Calculate the monthly payment for the cars in the following table by using the formula

$$M = \frac{iL}{1 - (1 + i)^{-60}}.$$

Which cars can you afford?

Year	Model	Price
1998	Chevy Malibu LS	$8,795
1999	Honda Civic LX	$11,995
2000	Mazda 626 LX	$10,995
2002	Saturn SL	$12,315

(b) Now that you have decided what cars you can afford, you realize you have not included the cost of gasoline. Assume the miles you drive are mostly highway, the average price for a gallon of gas is $1.43, and you travel about 13,000 miles a year.

Year	Model	MPG
1998	Chevy Malibu LS	25
1999	Honda Civic LX	40
2000	Mazda 626 LX	33
2002	Saturn SL	40

i. Determine the monthly cost of gasoline for each of the cars.

ii. What is the total monthly cost for each car (including both car payment and cost of gas)?

iii. Can you still afford the same cars? If not, which ones can you afford?

(c) Besides gas and car payments, you are required to have automoble insurance. The insurance rates listed below assume you've received two traffic tickets in the past year.[9]

    i. Determine the monthly cost for insurance for each of the cars.

    ii. What is the total monthly cost for each of the cars (including car payment, cost of gas, and insurance payments)?

    iii. Can you still afford the same cars? If not, which ones can you afford?

Year	Model	Cost of Insurance per Six Months
1998	Chevy Malibu LS	$1285.63
1999	Honda Civic LX	$1393.49
2000	Mazda 626 LX	$1431.55
2002	Saturn SL	$1153.79

---

[9] The insurance was also based on a driver who was a non-smoker, age 23, had a $500 PLPD deductible, and a $500 collision deductible, quoted by State Farm Insurance, July 2, 2002.

7

Linear Functions

We have looked at various ways to present graphical information. We will now look at various types of functions. We will begin by studying linear functions, the simplest and most common type of function. Common examples of linear functions include the conversion of feet to meters, dollars to euros, and temperature from the Fahrenheit scale to the Celsius scale. In this section, we will explore some of the important characteristics of linear functions and how to determine if a given function is linear.

Definitions

Table 7.1 is a representation of a linear function. Examine the relationship between the input and the output by filling in the blanks below.

TABLE 7.1
Tabular representation of a linear function.

Input	0	1	2	3	4
Output	1	3	5	7	9

- When the input changes from 0 to 1, the output changes from 1 to _____.
  When the input increases by 1 unit, the output changes by _____ units.
  The change in output divided by the change in input is _____.

- When the input changes from 0 to 2, the output changes from 1 to _____.
  When the input increases by 2 units, the output changes by _____ units.
  The change in output divided by the change in input is _____.

- When the input changes from 0 to 3, the output changes from 1 to _____.
  When the input increases by 3 units, the output changes by _____ units.
  The change in output divided by the change in input is _____.

- When the input changes from 0 to 4, the output changes from 1 to _____.
  When the input increases by 4 units, the output changes by _____ units.
  The change in output divided by the change in input is _____.

Notice the relationship. The change in output divided by the change in input is always the same, namely 2. This happens regardless of whether the input is changing by 1 unit, by 4 units, or by any other number of units. This will also happen if you start at 0, 1, or any other number. For this function, the change in output is always twice the change in input. The pattern is consistent: if the input changes by n, the output changes by $2n$. This constant rate of change defines a linear function. Thus, a **linear function** is a function where the ratio, $\frac{\text{change in output}}{\text{change in input}}$, is constant over the entire domain of the function.

Figure 7.1 is the graph of the points given in Table 7.1. Figure 7.1(a) contains just the points, while Figure 7.1(b) contains the points connected by a straight line. Do you see why this type of function is called linear?

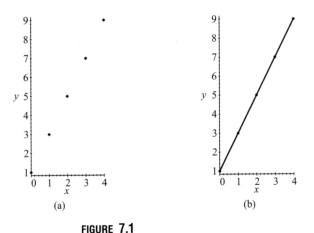

FIGURE 7.1
Graph of the points in Table 1.

Think about what it takes to define a linear function (or, equivalently, what it takes to draw a straight line). You need a starting point (somewhere to put your pencil) and you need to know how much the output changes when the input changes by 1 unit. For example, suppose you are asked to draw a line starting on the y-axis at the point $(0, 2)$ with the change in output being 3 times the change in input. You just need to plot the starting point and a second point and connect them to draw your line. You are given the starting point, $(0, 2)$, so you need to locate a second point. Since the change in output is 3 times the change in input, you can begin at $(0, 2)$, move to the right one unit and then move up three units to locate the second point. This produces $(0 + 1, 2 + 3) = (1, 5)$ as your second point. Connecting these two points with a line gives you the graph shown in Figure 7.2.

Note that these two pieces of information, where to start and the constant rate of change, are sufficient for drawing the graph of our linear function. This is what makes linear functions simple. You only need two pieces of information to completely define the

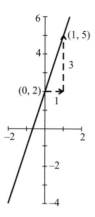

FIGURE 7.2
Graph of a linear function that has a starting point of $(0, 2)$ and whose change in output is 3 times the change in input.

function. There are formal definitions for these two items. The **y-intercept** is the point where the line crosses the y-axis (or, equivalently, the output when the input is zero). This can be thought of as our starting point. The **slope** is the constant rate of change or

$$\text{slope} = \frac{\text{change in output}}{\text{change in input}} = \frac{\text{change in } y}{\text{change in } x} = \frac{y_2 - y_1}{x_2 - x_1}$$

where (x_1, y_1) and (x_2, y_2) are any two points on the line. For the linear function given in Table 7.1, the y-intercept is 1 and the slope is 2. For the linear function given in Figure 7.2, the y-intercept is 2 and the slope is 3.

Let's look at another example. This time, draw the line with a y-intercept (or starting point) of 2 and a slope (or rate of change) of -3. A negative slope means that, for every change of one unit in the input, the output *decreases* by three units. To draw this graph, start at $(0, 2)$, but this time when you go one unit to the right (change the input by one), you go *down* three units (decrease the output by three). So the second point is

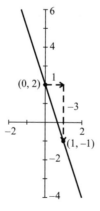

FIGURE 7.3
Graph of a linear function with a y-intercept of 2 and a slope of -3.

$(0 + 1, 2 - 3) = (1, -1)$. Connecting these two points creates the graph of this linear function. (See Figure 7.3.)

Symbolic Representations of Linear Functions

Remember that, for a function to be linear,

$$\text{the slope} = \frac{\text{change in output}}{\text{change in input}} = \frac{y_2 - y_1}{x_2 - x_1}$$

must be constant over the entire domain. Also remember that if we are given the y-intercept and the slope, we can draw the graph of the linear function. Similarly, if we are given the y-intercept and the slope, we can write a formula for the function. The y-intercept will again be our "starting point," i.e. the output when the input is zero. This means that $f(0) = y$-intercept. (See Figure 7.4.)

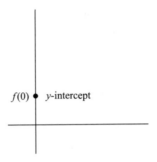

FIGURE 7.4
Starting point.

The slope tells us how much the output increases every time the input increases by one. Therefore, $f(1)$ will be where we started (the y-intercept) plus the slope. In symbols, $f(1) = \text{slope} + y$-intercept. (See Figure 7.5.)

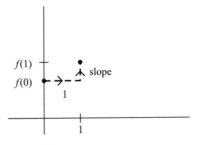

FIGURE 7.5
Finding a second point using the slope.

In Figure 7.5, the starting point is identified as the y-intercept. The second point, $f(1)$, is identified by starting at the y-intercept, moving horizontally one unit, and moving up (or down) the distance given by the slope. The remaining equations and pictures are a repetition of this process. In each case, you start where you left off, move one unit to the right horizontally, and move up (or down) the distance given by the slope.

$$f(2) = \text{slope} + f(1)$$
$$= \text{slope} + (\text{slope} + y\text{-intercept})$$
$$= (\text{slope} \times 2) + y\text{-intercept}$$

Figures 7.6 and 7.7 show how $f(2)$ and $f(3)$ are found using the previous point and the slope.

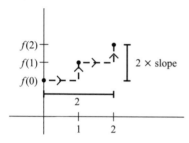

FIGURE 7.6

Finding a third point using the slope.

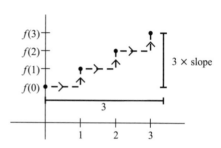

FIGURE 7.7

Finding a fourth point using the slope.

$$f(3) = \text{slope} + f(2)$$
$$= \text{slope} + \big((\text{slope} \times 2) + y\text{-intercept}\big)$$
$$= (\text{slope} \times 3) + y\text{-intercept}$$

Notice the pattern in our formula for finding $f(0)$, $f(1)$, $f(2)$, and $f(3)$. In each case, we started with the y-intercept and added x times the slope, where x was our input. Symbolically, this pattern is denoted as:

$$\text{output} = (\text{slope} \times \text{input}) + y\text{-intercept}$$

or

$$f(x) = (\text{slope} \times x) + y\text{-intercept}.$$

It is customary to use the letter x to represent the input, y to represent the output, m to represent the slope, and b to represent the y-intercept. Thus, the general equation for a line is

$$y = mx + b.$$

This equation, $y = mx + b$, is called the **slope-intercept** equation of a line.[1]

---

[1] There are other forms of the equation of a line, but the one we'll use in this text is the slope-intercept form.

It's easy to write the equation for a line if you know the slope and y-intercept. For example, the function given in Table 7.1 has the equation $y = 2x + 1$ since the slope is 2 and the y-intercept is 1. But what if you don't know the slope and y-intercept? Then how do you find the equation of the line? It happens that given any two points on a line, you can use a bit of algebra to find the slope, y-intercept, and therefore, the symbolic representation of the line that passes through those two points. This is why it is true that "two points determine a line." We'll illustrate this process by finding the equation of the line through the points $(1, 2)$ and $(5, 4)$.[2]

The slope of the line containing these two points is

$$m = \frac{y_2 - y_1}{x_2 - x_1} = \frac{4 - 2}{5 - 1} = \frac{2}{4} = \frac{1}{2}.$$

Now, in order to write the equation, we just need to know the y-intercept. You can figure out the y-intercept by reasoning it out in your head:

> I need to go to the left (backwards) one unit to get from the point $(1, 2)$ to the y-intercept. Since the slope is $\frac{1}{2}$, I need to go down half a unit (since I'm going to the left), so my y-intercept is $1\frac{1}{2}$. (See Figure 7.8.)

However, this reasoning can be quite confusing, particularly if the point you are using is not close to the y-axis or if the slope is not a simple number. Fortunately, algebra works well in this situation. Symbolically, you can find the y-intercept by using either of the given points and the slope. Start with the general equation for a line, $y = mx + b$, then substitute your slope for m, the coordinates of your point for x and y, and solve for b. For our example, we have taken the equation $y = mx + b$ and substituted our slope, $\frac{1}{2}$, for m; the x-coordinate of our point, 1, for x; and the y-coordinate of our point, 2, for y; obtaining:

$$2 = \left(\frac{1}{2} \times 1\right) + b$$

$$2 = \frac{1}{2} + b$$

$$\frac{3}{2} = b.$$

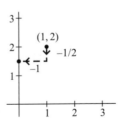

FIGURE 7.8
Finding the y-intercept using the point $(1, 2)$ and the slope.

---

[2] Symbolically, the two points you are given are typically referred to as (x_1, y_1) and (x_2, y_2).

So b, our y-intercept, is $\frac{3}{2}$, the same answer we obtained when we reasoned through the solution. The equation of the line containing the points $(1, 2)$ and $(5, 4)$ is $y = \frac{1}{2}x + \frac{3}{2}$.

The procedure illustrated in this example always works. Anytime you are given two points on a line and you want to find the equation for the line, you:

- Find the slope by computing

$$\frac{\text{change in } y}{\text{change in } x} = \frac{y_2 - y_1}{x_2 - x_1}.$$

- Substitute either one of your points and the slope you just found into the equation $y = mx + b$.

- Solve for b.

- Write the equation in the form $y = mx + b$.

Numerical Representations of Linear Functions

Suppose that a function is represented numerically rather than graphically or symbolically. In this case, how do we determine if the function is linear? Once again, we rely on the definition of linear functions, i.e., the function has constant slope over the entire domain. The symbol Δ (the upper-case Greek letter delta) is commonly used to denote change in a variable. We can therefore refer to the change in x as Δx (or "delta x") and the change in y as Δy (or "delta y"). $\Delta y/\Delta x$ is another way of writing the slope, m. Look at Table 7.2(a) and Table 7.2(b). One is a linear function and one is not. Can you tell which is which?

Looking at the slope, $\Delta y/\Delta x$, makes it easy to determine which one is the linear function. The function given by Table 7.2(a) has a constant slope of 2 over the entire domain, while the function given by Table 7.2(b) does not have a constant slope.

TABLE 7.2
Deciding if a table of numbers represents a linear function.

x	-2	-1	0	1	2
y	-4	-2	0	2	4
$\Delta y/\Delta x$		2/1	2/1	2/1	2/1

(a)

x	-2	-1	0	1	2
y	8	2	0	2	8
$\Delta y/\Delta x$		$-6/1$	$-2/1$	2/1	6/1

(b)

Remember, the definition of a linear function is that the slope is constant. If m denotes the slope, then one unit of change in the input will always produce m units of change in the output. In Table 7.2(a), as x increases by 1, y always increases by 2.

Suppose we wish to find an equation for the linear function given by Table 7.2(a). We have already found that the slope is 2. Remember that the y-intercept is the value of our linear function when the input is zero. Since Table 7.2(a) contains 0 as an input, we can see that the y-intercept is 0. So our equation is $y = 2x + 0$ or just $y = 2x$.

Example 1. Find the equation for the linear function given in Table 7.3.

TABLE 7.3

x	1	2	3	4	5
y	0	3/2	3	9/2	6
$\Delta y/\Delta x$		3/2	3/2	3/2	3/2

Solution. The slope is consistently $3/2$. We are not given the y-intercept in the table since zero is not an input. However, we can still find the equation of the line by using the procedure "find the equation of the line given two points." We can use *any* point in the table to substitute for x and y in the equation $y = mx + b$. We will use the point $(1, 0)$ since at least one of the two numbers is zero.[3] Substituting this point and our slope into the equation, we find that

$$0 = \left(\frac{3}{2} \times 1\right) + b.$$

Solving this, we obtain $b = -\frac{3}{2}$. Therefore, our equation is $y = \frac{3}{2}x - \frac{3}{2}$. ∎

BEWARE OF FALSE LINES!

Sometimes it is not easy to determine whether a function is linear when looking at a graph. For example, compare the two graphs in Figure 7.9. They both "look" like straight lines so one might conclude that they both are linear functions. However, Figure 7.9(a) is actually the graph of the function $y = \frac{20}{x}$ on the domain $19 \leq x \leq 20$. Figure 7.9(b) is the graph of the linear function $y = -0.05x + 2$ on the domain $19 \leq x \leq 20$.

The problem with just looking at the given portion of a graph and determining conclusively that it must be a linear function is that you are seldom seeing the entire function. It is therefore hard to know whether or not your function has a constant slope *everywhere* on its domain. The other problem is that the viewing window plays a huge role in determining the shape of the graph. The same two functions from Figure 7.9 are shown in Figure 7.10, but this time with a domain of $5 \leq x \leq 20$. Notice that Figure

---

[3] When given a choice, almost always go for the zero!

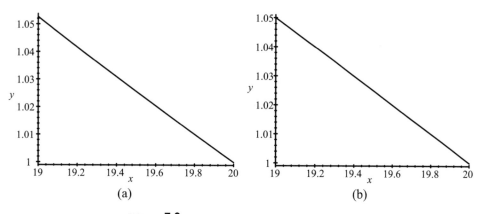

FIGURE 7.9
One of these is a linear function and one is not.

7.10(a) (the graph of $y = \frac{20}{x}$) no longer looks like a straight line while Figure 7.10(b) (the graph of $y = -0.05x + 2$) still does. What appeared to be a linear function when using one viewing window is obviously not a linear function when using another viewing window.

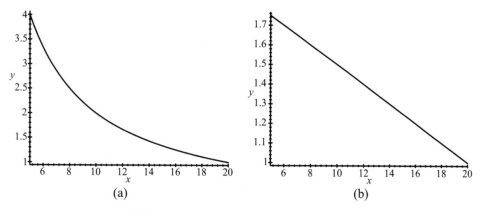

FIGURE 7.10
Same functions as in Figure 7.9, with a domain of $5 \le x \le 20$.

In fact, using a carefully chosen viewing window will cause almost any function to look linear.[4] You need to be careful in concluding that a graph of a "straight" line represents a linear function. It's always best to look at your function using a variety of viewing windows.

If you do conclude that your function is linear, we have already seen that it is easy to find its equation. For example, consider the function given in Figure 7.11. With this graph we cannot determine the y-intercept by just reading it off the graph because the

---

[4] Try it yourself. Use your graphing calculator and input some non-linear function. Then use the zoom-in function of your calculator over and over until the graph looks like a line.

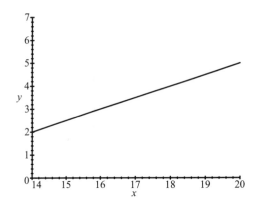

FIGURE **7.11**
Finding the equation of a linear function using a graph.

graph does not contain the y-axis. (The numbers on the x-axis begin with 14 rather than 0.) However, we can still determine the equation by estimating two points on the line. Using a ruler, the graph appears to contain the two points $(14, 2)$ and $(20, 5)$. So the slope is

$$m = \frac{y_2 - y_1}{x_2 - x_1} = \frac{(5 - 2)}{(20 - 14)} = \frac{3}{6} = \frac{1}{2}.$$

Substituting the point $(14, 2)$ and the slope, $\frac{1}{2}$, into the equation $y = mx + b$, we get

$$2 = \left(\frac{1}{2} \times 14 \right) + b.$$

Solving this, we find $b = -5$. So the equation of this line is $y = \frac{1}{2}x - 5$.

Conclusively determining that a function represented numerically is linear also has its difficulties. Again, this is because a linear function must have constant slope *everywhere* on its domain and your table will only have a finite number of points taken from the domain. This is especially a problem if you are working with a modeling function. You may have just a few points belonging to the function and might decide that a linear function best models your phenomena. So you find an equation representing your data and use this to predict what your phenomena will do in the future. However, if the function is not really linear (i.e., doesn't always have constant slope) your prediction will be wrong and could cause significant problems.

For example, suppose you collect the data shown in Table 7.4.

TABLE **7.4**
Three data points.

x	0	1	2
$f(x)$	0	2	4

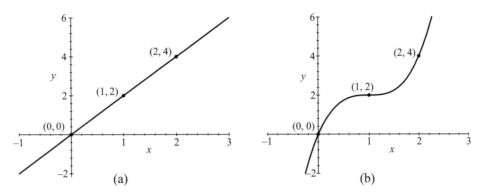

FIGURE 7.12
Many different functions contain the three points given in Table 7.4.

The linear equation which fits this data is $f(x) = 2x$. (See Figure 7.12(a).) However, the equation $f(x) = 2x^3 - 6x^2 + 6x$ or even something as complicated as

$$f(x) = \frac{4}{\sqrt{3}} \sin\left(\frac{\pi(x-1)}{3}\right) + 2$$

also fits the data. (See Figure 7.12(b).) Knowing which of these is the best equation for your model depends on collecting more data points and gaining a deeper understanding of the phenomenon producing the data. Further research is needed before assuming linearity!

Sometimes a function that actually is linear may be written symbolically in another format. This may make it difficult to initially see that it is a linear function. Remember that the equation for a linear function is $y = mx + b$. Any function which is linear can be written in this form. You may have to algebraically solve for y (or whatever letter represents the output) before deciding if your function is linear. Consider the equation

$$\frac{y-4}{x} = 7.$$

This does not "look" like the equation of a line, but if we solve for y (by multiplying both sides by x and then adding 4), we get

$$y = 7x + 4$$

which is the equation for a linear function with slope 7 and y-intercept 4.[5] Now consider the equation

$$\frac{y-4}{x} = 7 + x.$$

When we solve this for y by multiplying both sides by x and then adding 4, we get

$$y = 7x + x^2 + 4.$$

---

[5] The original equation $\frac{y-4}{x} = 7$, excludes the input $x = 0$ because of division by zero. So, technically, this is a linear function which is not defined for $x = 0$.

This is not a linear function since the equation has an x^2 term. To be safe, always solve the equation for y (or whatever letter represents the output) and see if it looks like $y = mx + b$ before deciding if the function is linear.

Verbal Representations of Linear Functions

Many situations are best modeled with a linear function. Not surprisingly, any situation with a *constant* rate of change can be described by a linear function. For example, if you travel at a *constant* velocity of 60 mph, then the distance, d (in miles), that you travel is represented by the linear function

$$d = 60t,$$

where t is time (in hours). When we have driven for 0 hours ($t = 0$), we have gone 0 miles ($d = 0$). So the y-intercept is zero. The slope is 60 since the constant rate of change (i.e., velocity) is 60 mph. That is, for every 1 hour that you drive, your output (distance) increases by 60 miles.

The key to determining if a verbal representation or a physical situation is best modeled by a linear function is no different than for graphical or numerical representations: you need to determine if there is a constant rate of change. Once again, you can find the equation by determining two pieces of information, the slope and the y-intercept. In a physical (or verbal) situation, the y-intercept will be your starting point (i.e. your output when your input is zero) and the slope will be your constant rate of change. [Note: Since slope is a *rate* of change, the units of slope will always be *something* per *something* such as miles per hour, wages per hour, cost per month.]

Example 2. Suppose you call a plumber to repair some problem at your house. He charges \$25 to show up at your house and \$50 per hour. Let C be the function whose input is n, the number of hours the plumber works and whose output is $C(n)$, the *total* cost. Explain why this is a linear function and find an equation for C.

Solution. This is a linear function since the rate of change (in this case, the charge per hour) is constant. To find the equation, we need to find the slope and the y-intercept. The slope, or rate of change, is the \$50 charge per hour. The y-intercept is the cost for the plumber to show up or the starting cost which is \$25. So the equation for this function is

$$C(n) = 50n + 25$$

where $C(n)$ is the total cost (in dollars) and n is the number of hours the plumber works.
■

Linear functions are often the basis for converting units of measurement. We measure many things, e.g., length, weight, time, velocity, sound levels, intelligence. Each of these can be measured in a variety of units. Length, for example, can be measured in inches, feet, meters or miles. Converting from one system of measurement to another almost always uses a linear function. This is because the rate of change for measurement is *constant*, e.g., there are always 12 inches per foot, 2.54 centimeters per inch, or 5280 feet per mile. These types of conversions are often given verbally by such statements as

"There are 12 inches in a foot" or "There are 5280 feet in a mile." Suppose you want to write one of these statements symbolically. For example, think about how you convert from feet to inches. To find how many inches are in 2 feet, you could write

$$2 \text{ feet} \times \frac{12 \text{ inches}}{1 \text{ foot}} = 24 \text{ inches}.$$

[Note the cancellation of units to produce the correct unit for the output. The "feet" cancels the "foot" in the denominator, leaving "inches" as the unit.] To find how many inches are in 5 feet, you could write

$$5 \text{ feet} \times \frac{12 \text{ inches}}{1 \text{ foot}} = 60 \text{ inches}.$$

To find how many inches are in f feet, you could write

$$f \text{ feet} \times \frac{12 \text{ inches}}{1 \text{ foot}} = 12f \text{ inches}.$$

This process can written symbolically as

$$\text{number of feet} \times 12 = \text{number of inches}$$
$$\text{or}$$
$$12f = i$$

where f is the number of feet and i is the number of inches. Notice that, no matter how many feet you have in your measurement, changing the number of feet by one means increasing the number of inches by twelve. The rate of change of this function is constant, which is the definition of a linear function. Notice that the slope, 12, is the conversion factor, i.e., there are 12 inches *per* foot. The y-intercept is zero since if the input (feet) is zero, then the output (inches) is also zero.

Let's now look at the linear function used for converting from quarts to gallons. There are four quarts in one gallon. So,

$$\text{number of quarts} \times \frac{1 \text{ gallon}}{4 \text{ quarts}} = \text{number of gallons}$$
$$\text{or}$$
$$\frac{1}{4}q = g.$$

Notice, again, the cancellation of units to produce the correct unit for the output.[6] Also notice that the y-intercept is again zero. This is because zero in the first unit of measurement corresponds to zero in the second unit of measurement. Zero quarts means zero gallons. Zero feet means zero inches. Will the y-intercept for a linear conversion function always be zero?

The answer is "no." A familiar example of this is temperature. Zero degrees Celsius is cold, but it does not mean that there is no temperature or heat present.[7] You may be aware

---

[6] Using cancellation of units is a good way of deciding if you want (1 gallon)/(4 quarts) or (4 quarts)/(1 gallon).

[7] There is a scale for temperature in which $0°$ represents the complete absence of heat. Zero on the Kelvin scale is $-273°\text{C}$ and is often referred to as absolute zero.

that 0°C will actually feel quite warm when compared to 0°F! Zero degrees Celsius equals 32 degrees Fahrenheit, the freezing point of water. This means that the linear function that converts from temperature in the Celsius scale to the Fahrenheit scale must contain the point $(0, 32)$, i.e., have a y-intercept of 32. To find the conversion equation for changing from Celsius to Fahrenheit, we need one more point (remember, "two points determine a line"). Knowing that water boils at 100°C and at 212°F produces our second point. Using the two points, $(0, 32)$ and $(100, 212)$, we find that the slope is $\frac{212-32}{100-0} = 1.8$. So the conversion equation is the linear equation F $= 1.8$C $+ 32$. The y-intercept of 32 is because 0°C $=$ 32°F (an input of zero gives an output of 32). The slope is 1.8 because a change of 1 degree Celsius is equivalent to a change of 1.8 degrees Fahrenheit.

Historical Note.[8]   Many units of length have their origins with human anatomy. For example, an inch was once defined as the width of a person's thumb. A yard was defined as the length from a person's nose to the tip of their finger of their outstretched arm. Some people still measure cloth this way. Measuring length by using one's body naturally varied from person to person. Through the years, people tried to account for this variation by finding averages. In the 16th century, a German named Master Koebel wrote how to produce the "right and lawful" length for the standard foot for surveying. He suggested that the surveyor should stand outside a church door on Sunday. After the service, as people were filing out of the church, the surveyor was to ask the first 16 men out the door (a semi-random selection) to stand in a line with their left feet one behind another. This would produce a line 16 "feet" long. Since $16 = 2^4$, a string that was cut to this sixteen foot length would just have to be folded in half four times to produce a length of one foot.

Application: Gasoline Prices

It has been claimed that gasoline prices are a linear function of the octane level of the gasoline. To test this assumption, the octane level and gasoline prices were collected from three different gasoline stations as shown in Table 7.5.[9]

TABLE 7.5
Octane level and gasoline prices at three gasoline stations

Octane Level	Citgo	Shell	Speedway
87	1.40	1.42	1.41
89	1.50	1.52	1.51
93	1.60	1.62	1.61

If the gasoline prices are a linear function of the octane level, the slope will be constant. In other words,

$$\frac{y_2 - y_1}{x_2 - x_1}$$

will have the same value for any two points, where y is the price of gasoline, and x is the octane level.

---

[8]   Klein, H. Arthur, *The World of Measurements*, New York: Simon and Schuster, 1974, p. 66.
[9]   Gas prices in Holland, MI on July 3, 2002.

Example 3. Are the Citgo gasoline prices a linear function of the octane level?

Solution. Since there are three points, we have to compare two slopes: one from 87 to 89 and one from 89 to 93. The slope of the line between the first two octane levels for Citgo is $\frac{1.50-1.40}{89-87} = 0.05$. In order for this function to be linear, the other slope must also be 0.05. However, the slope between the 89 and 93 octane levels is $\frac{1.60-1.50}{93-89} = 0.025$. There is an equal change in the output (i.e., \$0.10), but there is not an equal change in the input (2 versus 4). This means that the Citgo data do not exhibit a linear relationship. ■

   Figure 7.13 shows a graph for all three gasoline stations. As you can see, none of the gasoline stations shows a linear relationship between octane level and gasoline prices. What is interesting is that the slopes are the same between the same octane levels. All three gasoline stations show a slope of 0.05 between the 87 and 89 octane level and a slope of 0.025 between the 89 and 93 octane level. Each of these graphs has the same shape but have been shifted by adding a constant to the output. If $y = c(g)$ is the function for the Citgo data, then the Speedway function is $y = c(g) + 0.01$ and the Shell function is $y = c(g) + 0.02$.

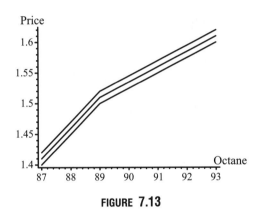

FIGURE 7.13

Reading Questions for Linear Functions

1. What is the definition of a linear function?

2. What two pieces of information determine a line?

3. A graph containing two points is to the right.

   (a) What are the coordinates of A?

   (b) What is the equation for this line?

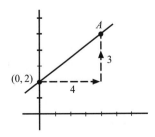

4. Suppose $y = f(x)$ is a linear function whose slope is 2. If $f(1) = 1$, what is $f(2)$? $f(3)$? $f(10)$?

5. Let $y = f(x)$ be a linear function whose y-intercept is -1 and whose output changes by 2 when the input changes by 1. What is $f(2)$?

6. Suppose $f(x) = 3x - 1$.

   (a) How do you know f is a linear function?

   (b) What is the slope?

   (c) What is the y-intercept?

   (d) What is $f(0)$?

   (e) What is $f(1)$?

7. Sketch each of the following. Label at least three points on your graph.

   (a) A linear function with y-intercept 0 and slope $\frac{1}{3}$.

   (b) A linear function with y-intercept 1 and slope -2.

8. Let A, B, and C be three points such that the slope between A and B equals the slope between B and C. What, if anything, do you know about the slope between A and C? Explain.

9. Find the equation of the line for each of the following.

   (a) The y-intercept is 5 and the slope is $-\frac{1}{4}$.

   (b) It contains the two points $(0, 10)$ and $(1, 11)$.

   (c)
Input	-2	-1	0	1	2
Output	0	2	4	6	8

   (d)

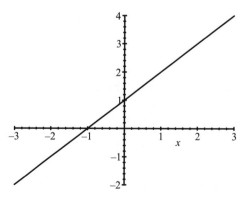

10. Determine if each of the following tables represent a linear function. If so, find an equation for the line. If not, **explain** why.

   (a)
Input	-2	-1	0	1	2
Output	3	0	-3	-6	-9

(b)

Input	−2	−1	1	2
Output	−4	0	8	12

(c)

Input	−2	−1	0	1	2
Output	12	10	8	10	12

(d)

Input	1	2	3	4	5
Output	−6	−4	−2	0	2

11. For the tables in Question 10 that represented lines, which ones included the y-intercept?

12. Explain some of the pitfalls you might encounter when determining if a graph represents a linear function.

13. Determine whether or not each of the following represents a linear function.

   (a) $\dfrac{y-1}{x+2} = 4$

   (b) $y - 4 = 1.2(x + 3)$

   (c) $\dfrac{x+3}{y} = 16$

14. A calling card offers a price of $0.23 cents per minute at any time during the day with a monthly service charge of $1.00.

   (a) Write a linear equation (including the $1.00 monthly service charge) where the input is minutes and the output is total cost.

   (b) Describe the practical interpretation of the y-intercept.

   (c) Using your equation from part(a), calculate the total cost for a month if you talk, on average, for 20 minutes a day. Assume the month is thirty days long.

15. Graph the following functions on your calculator and zoom in on the point $(1, 1)$ so that they appear linear. Give the viewing window you used for each.

   (a) $y = x^2$

   (b) $y = \sqrt{x}$

   (c) $y = \frac{1}{x^2}$

16. Which is the longer run in a track meet: the one mile run or the 1600 meter run? Justify your answer. (1 meter $\approx$ 39.37 inches)

17. Give the linear equation that will convert from the first unit to the second.

   (a) centimeters to inches (2.54 centimeters = 1 inch)

   (b) yard to inches (1 yard = 36 inches)

   (c) degrees Fahrenheit to degrees Celsius. [Hint: Use the points $(32, 0)$ and $(212, 100)$.]

18. We are interested in converting from miles per hour to feet per second.

    (a) Convert 70 mph to feet per second.

    (b) Write a linear conversion function whose input is the velocity in miles per hour and whose output is the velocity in feet per second.

19. In Example 3, we found that gasoline price was not a linear function of the octane level. For the data shown in the table below, what would the price of 93 Octane level gasoline be at each of the three gasoline stations if the slope between the 89 and 93 octane levels is the same as the slope between the 87 and 89 octane levels?vadjust

Octane Level	Citgo	Shell	Speedway
87	1.40	1.42	1.41
89	1.50	1.52	1.51
93	1.60	1.62	1.61

Linear Functions: Activities and Class Exercises

1. **Producing Linear Equations with a Motion Detector.** Use a motion detector to answer the following questions.

    (a) Have a member of your group walk in front of the motion detector to produce each of the following graphs. Describe how the person walked in order to obtain the appropriate graph. Be sure to record the y-intercept and one other point for each graph (you'll need this in part(b)).

        i. A line with slope between 1.8 and 2.2.

        ii. A line with slope between 3.8 and 4.2.

        iii. A line with slope between -2.2 and -1.8.

    (b) For each graph in part(a), find the y-intercept and the slope.

    (c) What is the physical significance of the y-intercept?

    (d) What is the physical significance of the slope?

2. **Currency Exchange Rates.** The currency exchange rates on July 3, 2002 are given below.[10] The column represents the currency you are starting with and the row represents the currency you are exchanging to. For example, 1 U.S. dollar converts to 1.371 Australian dollars.

---

[10] *Yahoo! Finance* — "Currency Conversion," 12 March 2004, <http://quote.yahoo.com/m3>.

Currency	U.S.$	Yen	Euro	Can $	U.K. Pound	Aust $	SFranc
U.S. $	1	0.008988	1.222	0.7487	1.797	0.7295	0.7788
Yen	111.3	1	136	83.3	200	81.16	86.65
Euro	0.818	0.007352	1	0.6124	1.47	0.5967	0.6371
Can $	1.336	0.01201	1.633	1	2.401	0.9744	1.04
U.K. Pound	0.5564	0.005001	0.6801	0.4165	1	0.4059	0.4333
Aust $	1.371	0.01232	1.676	1.026	2.464	1	1.068
SFranc	1.284	0.01154	1.57	0.9613	2.308	0.9367	1

(a) Answer the following questions which involve converting from dollars to yen.

    i. Fill out the following table.

U.S. dollars	Japanese Yen
$300	
$400	
$500	

    ii. For every increase of $100 in the input (U.S. dollars), the output (Japanese yen) increased by how much?

    iii. The function which converts from U.S. dollars to Japanese yen is a linear function. How do you know this?

    iv. Give a symbolic formula for the function whose input is U.S. dollars and whose output is Japanese yen.

    v. What is the slope of your function? What is the meaning of the slope (in terms of dollars and yen)?

    vi. What is the y-intercept of your function? What is the meaning of the y-intercept (in terms of dollars and yen)?

(b) Answer the following questions which convert from Japanese Yen to U.S. dollars.

    i. Fill out the following table.

Japanese Yen	U.S. Dollars
3000	
4000	
5000	

    ii. For every increase of 1000 in the input (Japanese yen), the output (U.S. dollars) increased by how much?

    iii. The function which converts from Japanese yen to U.S. dollars is a linear function. How do you know this?

    iv. Give a symbolic formula for the function whose input is Japanese yen and whose output is U.S. dollars.

    v. What is the slope of your function? What is the meaning of the slope (in terms of yen and dollars)?

    vi. What is the y-intercept of your function? What is the meaning of the y-intercept (in terms of yen and dollars)?

(c) Compare the slope of the linear function that converts from dollars to yen (part (a)v.) to the slope of the linear function that converts from yen to dollars (part (b)v.). What is the relationship between these two numbers? [Hint: Multiply the two numbers together.]

(d) Why would you expect this relationship to be true?

3. **Electric Bill.** Below is information from an electric bill for June 1997 from Consumers Power. Notice that there are three different rates. Let $e(k)$ be the function whose input is kilowatt/hours (KWH) and whose output is the cost of the electricity.

First 300 KWH × $0.090906	$27.27
Next 300 KWH × $0.076206	$22.86
Next 49 KWH × $0.090906	$4.45

649 KWH TOTAL	Total Electric	$54.58
Michigan Sales Tax		$2.18

(a) Explain why $e(k)$ is a piece-wise function.

(b) Let $e_1(k)$ be the function where the domain is restricted to the first 300 KWH hours.

    i. Explain why $e_1(k)$ is a linear function.

    ii. Find an equation for $e_1(k)$ and sketch it by hand for the appropriate domain.

    iii. Explain the meaning of the slope and y-intercept for $e_1(k)$ in terms of KWH and cost of electricity.

(c) Let $e_2(k)$ be the function where the domain is restricted to between 300 and 600 total kilowatt hours.

    i. Explain why $e_2(k)$ is a linear function.

    ii. Find an equation for $e_2(k)$.

    iii. Compute $e_2(300)$. Check to make sure that this equals $e_1(300)$.

    iv. Sketch $e_2(k)$ for the appropriate domain on the same graph as $e_1(k)$. [Note: The two lines should "connect" at 300.]

    v. Explain the meaning of the slope for $e_2(k)$ in terms of KWH and cost of electricity.

(d) Let $e_3(k)$ be the function where the domain is restricted to over 600 kilowatt hours.

    i. Explain why $e_3(k)$ is also a linear function.

    ii. Find an equation for $e_3(k)$.

    iii. Compute $e_3(600)$. Check to make sure that this equals $e_2(600)$.

    iv. Sketch $e_3(k)$ for the appropriate domain on the same graph as $e_1(k)$ and $e_2(k)$. [Note: Again, your lines should "connect."]

    v. Explain the meaning of the slope for $e_3(k)$ in terms of KWH and cost of electricity.

4. **Miles Per Gallon.** Owning a vehicle can be quite costly when all the factors, such as car payment, gasoline, maintenance, and insurance are included. The cost of gasoline is related to the miles per gallons averaged by the car. If you can improve this, the cost of gasoline will decrease (assuming you continue to drive the same number of miles).

(a) Suppose you intend to buy either a 1998 Chevy Malibu LS or a 1999 Honda Civic LX. The Malibu costs $8,795 while the Honda costs $11,995. However, the Civic gets 40 mpg while the Malibu only gets 25 mpg. Assume the cost of gasoline is $1.45 a gallon.

    i. Write linear equations for the Malibu and the Civic where the input is miles and the output is the total cost for owning and operating the vehicle. Include the initial cost of buying the vehicle but ignore maintenance and insurance costs.

    ii. Graph the equations on the same axis.

        A. Find the point of intersection.

        B. What is the physical meaning of the intersection point? Write your answer in a complete sentence.

    iii. Suppose your commute to work was forty miles round trip and you worked five days a week on average. Also assume you drove, on average, an additional 10 miles per day each day of the week. How long (in months) would it take before the Civic becomes the more economical vehicle? Give your answer in months.

(b) Street and Performance Electronics has developed a product that claims to increase your miles per gallon by up to four miles.[11] Suppose your car is currently getting 21 mpg and the device costs $137 dollars.

    i. Write a linear equation for your vehicle that represents the total cost of gasoline. Continue to assume gas costs $1.45 a gallon. Also, write a linear

---

[11] *Street & Performance Electronics*, 12 March 2004,
<http://www.streetandperformanceelectronics.com/lead.htm>.

equation for your vehicle (including the initial cost of the device) assuming the device has been installed, where the input is miles and the output is cost.

ii. Find the intersection of the two lines.

iii. What is the physical meaning of the intersection point? Give your answer in a complete sentence.

iv. How long would it take for the product to be economically worthwhile if, on average, you drive 13,000 miles a year?

8

Regression and Correlation

In the section *Displaying Data*, we looked at different types of graphs used to display single-variable data and two-variable data. In the section *Describing Data*, we looked at ways that numbers, such as mean and median, can describe single-variable data. In this section, we will look at how two-variable data can be described using both numbers and linear equations. We will see how to describe a linear relationship in a scatterplot using a line of best fit and its correlation.

Linear Relationships

Scatterplots are used to display relationships for two-variable data, many of which are approximately linear. For example, the average weight of an American infant in kilograms given the infant's age in months is shown in the table in Figure 8.1.[1] When these data are graphed, the pattern appears to be almost, but not quite, linear. Since these data are not actually linear, no one line will pass through every point. However, we can *fit a line to this data* by drawing a line that is close to all the points. We can then use the equation of this line as a symbolic description of the data.

A linear equation that fits the infant weight data will have the form

$$\text{weight} = m \times \text{age} + b,$$

where m is the slope of the line and b is the y-intercept. The slope represents the increase in weight each month and the y-intercept represents the weight at birth (0 months old). The linear equation that best fits the infant weight data given in the table in Figure 8.1 is[2]

$$\text{weight} = 0.77 \times \text{age} + 3.43.$$

---

[1] *NCHS*-"United States Clinical Growth Charts," 15 March 2004, <http://www.cdc.gov/nchs/about/major/nhanes/growthcharts/clinical_charts.htm>.

[2] We found this equation using linear regression on a TI-83 graphing calculator.

Age (months)	Weight (kilograms)
0	3.2
1	4.2
2	5.1
3	5.9
4	6.7
5	7.3
6	7.8

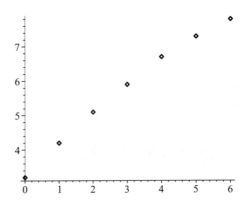

FIGURE 8.1
The average weight of infants, in kilograms, for the first six months in a table and a scatterplot.

This can also be written as

$$y = 0.77x + 3.43.$$

If this equation gave the exact connection between age and weight (i.e., if the data points were all on this line), it would imply that the average weight of an infant at birth would be 3.43 kilograms and the infant's weight would increase by approximately 0.77 kilograms each month. From the table in Figure 1 we can see that the average weight of a newborn is not 3.43 kilograms, but is actually 3.2 kilograms and that the weight increased by 1 kilogram, not 0.77, when the newborn was one month old. However, we can see that the linear equation closely approximates our data by drawing the line in the same graph as the scatterplot. (See Figure 8.2.) Notice that the line comes close to each point although it only appears to go through two of them. Having a linear equation that describes the pattern of the data allows us to communicate information in the compact form of a single equation.

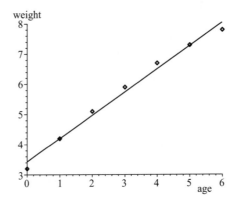

FIGURE 8.2
A scatterplot of the average weight of infants, in kilograms, for the first six months along with the line that best fits the data.

Now that we have an equation describing the relationship between age and the weight of infants, we can use it to predict an infant's weight given the infant's age. For example, to predict the weight of an infant that is 2.5 months old, we just substitute 2.5 into our equation to find

$$\text{weight} = 0.77 \times 2.5 + 3.43 \approx 5.4 \text{ kilograms}.$$

Since the data points in Figure 8.2 lie close to the line, we know that 5.4 kilograms is a fairly good estimate for the average weight of a 2.5 month-old infant. A word of warning, however. While our equation does a good job of predicting the weight of infants from 0 to 6 months, it does not necessarily do a good job of predicting the weight of older children. For example, using our equation to predict the weight of a 36 month-old child produces

$$\text{weight} = 0.77 \times 36 + 3.43 \approx 31.2 \text{ kilograms}.$$

Our prediction of 31.2 kilograms $\approx$ 69 pounds is not accurate at all. This is almost twice the weight of an average 36 month-old! This is because, while a newborn infant may gain about 0.77 kilograms a month for the first few months, this rate of weight gain soon slows down. In general, using the equation for the line of best fit to make predictions ONLY works well when your input is close to the inputs from your data points.

Least-Squares Regression

In the section *Displaying Data*, we looked at a scatterplot of students' heights and arm spans. Since we were seeing if a person's arm span is about the same as his or her height, we also included the line $y = x$ in the graph of our scatterplot. That scatterplot and the line $y = x$ are repeated in Figure 8.3. The line $y = x$ seems to fit the data, but is there a line that fits better? The answer to this question is yes.

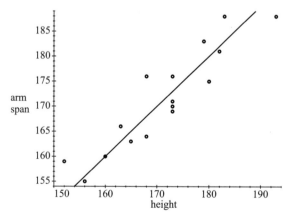

FIGURE **8.3**
A scatterplot of students' heights (in centimeters) and arm spans (in centimeters) with the line $y = x$.

TABLE **8.1**

Heights and arm spans (in centimeters) of Hope College students.

Height	152	160	165	168	173	173	180	183
Actual Arm Span	159	160	163	164	170	176	175	188
Predicted Arm Span using the line	153.59	160.55	164.90	167.51	171.86	171.86	177.95	180.56
Error	5.41	0.55	−1.90	−3.51	−1.86	4.14	−2.95	7.44

The most common method of determining the equation of the line of best fit is the *method of least-squares*. This method involves finding the difference between the predicted output from the line and the actual output from the data points and finding the equation which minimizes a combination of these differences. The line that does this is referred to as the least-squares regression line or simply the regression line. Formally, a **least-squares regression line** is a line that best fits two-variable data by minimizing the sum of the squares of the vertical distances between the points in the scatterplot and the line. To illustrate this, look at Table 8.1.

This is part of the height and arm span data graphed in Figure 8.3. We found, using a calculator, that the least-squares regression line for these data is

$$\text{arm span} = 0.87 \times \text{height} + 21.35.$$

The last row of Table 1 gives the error of our line, i.e., the difference between the actual output (using the data points) and the predicted output (using the equation of the line). The sum of the squares of all these errors for the least-squares regression line will be smaller than the sum we would get using any other line. This is why the least-squares regression line is often called the line of best fit.

The error between actual outputs (using the data points) and predicted outputs (using the equation of the line) is illustrated in Figure 8.4. This figure shows the data points for the heights and arm spans from Table 8.1 (represented by dots) along with the least-squares

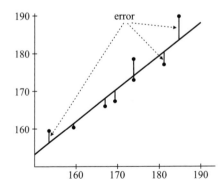

FIGURE **8.4**

The vertical distance from a point to a line that fits the data is the error. For a regression line, this error is minimized.

regression line, $y = 0.87x + 21.35$. In this figure, the lengths of the vertical lines between the points and the regression line represent the error. In fitting a line to the data, our goal is to make these vertical lengths as small as possible. The method of least-squares does this by minimizing the sum of the squares of these distances.

The equation for the least-squares regression line can be found using complicated formulas. However, graphing calculators and software packages such as Excel simplify this process.

Correlation

The two examples of regression lines that we have looked at so far do a reasonably good job of predicting values for inputs close to our data points. Both the infant weight equation and the height and arm span equation have predicted outputs that are close to the actual outputs. However, some regression equations do not give good results. For example, in a statistics class, students were asked to give their mother's and father's heights. We wanted to see if the mother's height could be predicted given the father's height. A scatterplot of these results along with the least-squares regression line is shown in Figure 8.5.

You can see that the regression equation will not give accurate predictions since many of the data points are not very close to the line. This is because the data does not follow a linear pattern. To measure how well a regression line fits a set of data, we use a number called correlation. **Correlation** measures both the strength and direction of a linear relationship between two numerical variables. The height and arm span data in Figure 8.3 is said to be highly correlated while the mother and father height data in Figure 8.5 is not.

Correlation is usually denoted by the letter r. Correlation is always a number between -1 and 1. Symbolically, we write $-1 \leq r \leq 1$. If the correlation is close to -1 or 1, then the data points are highly correlated, i.e., the plotted data points lie very close to a line. In fact, if $r = 1$ or $r = -1$, the data all lie on a straight line and the regression

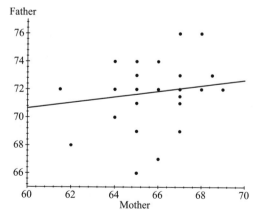

FIGURE 8.5

A scatterplot of the heights of students' mothers and fathers along with the regression line.

line perfectly predicts the outputs. If the correlation is close to 0, then the data points have little correlation, i.e., the plotted data points do not follow a linear pattern. If the correlation is positive, then there is a positive association between the two variables and the regression line will have a positive slope. This means that as one variable increases, the other also increases (such as height and arm span). If the correlation is negative, then there is a negative association between the two variables and the regression line will have a negative slope. This means that as one variable increases, the other decreases. An example of this is the relationship between the age of a relatively new car and its value.

The correlation of the height and arm span data shown in Table 8.1 and Figure 8.3 is 0.90 while the correlation of the father and mother height data in Figure 8.5 is 0.17. This means that the height and arm span data is highly correlated and has a positive association, i.e., people who are taller have longer arm spans. The father and mother height data has a very low correlation. This means that one parent's height is not a good predictor of the height of the other parent.

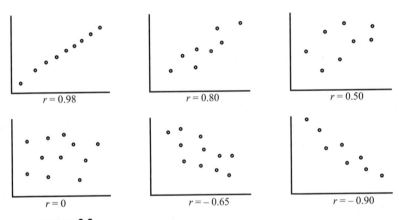

FIGURE 8.6
Examples of various scatterplots and their corresponding correlation.

You can often approximate the correlation simply by looking at a scatterplot of the data if the axes are scaled appropriately. Figure 8.6 shows a number of different scatterplots with corresponding correlations. As with the least-squares regression equation, there are rather complicated formulas to determine the correlation. However, we will rely on graphing calculators or software packages such as Excel to compute these numbers.

Correlation is often confused with causation. However, two variables can be highly correlated without one necessarily causing the other. For example, if you were studying the mathematical ability of elementary school students, you would probably find a high positive correlation between a child's shoe size and his or her ability in mathematics. This does not mean that the growth of a child's foot causes the child to become better at mathematics. It means that as children grow older, their feet are growing longer and at the same time they are learning more mathematics in school. If two variables are highly correlated, it just means that there is a relationship between them. It does not mean that one necessarily causes the other.

Application: Carbon Emissions from Fossil Energy Consumption

Table 8.2 gives carbon emissions from fossil energy consumption for the commercial sector in the United States.[3]

A scatterplot of the data is shown in Figure 8.7(a). We measured time as years since 1992. So 1992 is represented by 0, 1993 is represented by 1, etc. These points seem to portray a linear relationship. The least-squares regression line for this data is $y = 7.25x + 207$, with correlation $r = 0.98$.

TABLE **8.2**

Carbon emissions from the U.S. commercial sector from 1992 to 2001.

Year	1992	1993	1994	1995	1996	1997	1998	1999	2000	2001
Carbon Emissions (Million Metric Tons)	211	217	220	225	233	245	250	252	264	280

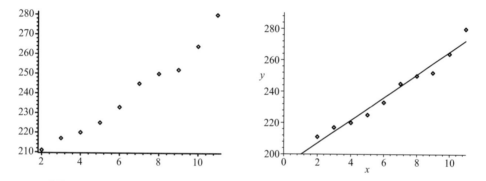

FIGURE **8.7**

(a) shows a scatterplot of the carbon emissions data. (b) shows the scatterplot with the least-squares regression line.

Example 1.   What is the physical interpretation of the slope and y-intercept of the least-squares regression line obtained from the data in Table 8.2?

Solution.   The y-intercept is 207, which means that at time 0 (the year 1992), approximately 207 million metric tons of carbon were emitted from fossil energy consumptions. This is very close to the actual 211 million metric tons emitted in 1992 given in Table 8.2. The line has a slope of 7.25 which means the amount of carbon emissions increased an additional 7.25 million metric tons each year. ∎

Looking at this trend, it is easy to see why environmental groups are becoming increasingly concerned with the need for the government to set higher standards for the commercial sector. If this pattern continues, carbon emissions just from the commercial sector will be 410.1 million metric tons in the year 2020.

---

[3] *Energy Information Administration*—"Flash Estimate," 15 March 2004, <www.eia.doe.gov/oiaf/1605/flash/flash.html>.

Reading Questions for Regression and Correlation

1. Suppose a baby boy's birth weight was 8.25 pounds and he gained 2.1 pounds each month for the first few months of his life. The function describing his weight is linear and could be written as

$$y = 2.1x + 8.25,$$

where x is his age in months and y is his weight in pounds.

   (a) Determine the baby's weight after 3 months.

   (b) Determine the child's weight after 36 months. Is your answer reasonable? Explain.

   (c) If the baby's birth weight was 8.5 pounds instead of 8.25, how would the function change?

   (d) If the slope of the function were 2.0, what would this mean?

2. The following is the graph of a scatterplot and its regression line.

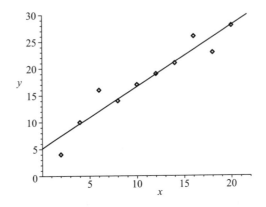

   Which of the following is the equation of the regression line? For each answer you do *not* choose, explain why.

   (a) $y = 5.13x + 1.15$

   (b) $y = 1.15x + 5.13$

   (c) $y = 10.3x + 5.13$

   (d) $y = -2.1x + 5.13$

3. Suppose you have a scatterplot and you draw in a line that you think nicely fits the data. How is the least-squares regression line different from the line you drew?

4. Can you ever have a least-squares regression line where all the data points are below the line? Explain.

5. What does correlation measure?

6. Match the following correlations with the corresponding scatterplot.

(i) $r = 0.6$　(ii) $r = 0.98$　(ii) $r = 0.85$　(iv) $r = 0$　(v) $r = -0.85$　(vi) $r = -0.6$

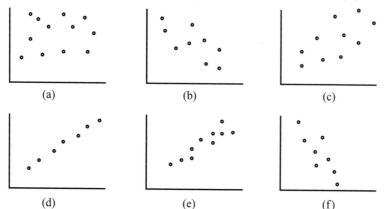

7. Just because two variables are highly correlated does not mean that one necessarily causes the other. There may be something else that is causing one or both events.

   (a) There is a high correlation between a child's height and his or her ability in mathematics. Does this mean that a child's physical growth is causing the increased ability in math? Explain.

   (b) There is a high correlation between the outside temperature and the number of colds people get. Does this mean that the number of colds people get causes the temperature to decrease? Explain.

   (c) There is a high correlation between the number of drinks someone has at a bar and incidents of lung cancer. Does this mean that drinking causes lung cancer? Explain.

8. Suppose a friend is telling you about a study in a popular magazine. The friend says the study reported a correlation of zero between the amount of coffee a person drank as a child and their height as an adult. Your friend interprets this to mean that the more coffee a child drinks, the shorter he or she will be as an adult. In other words, according to your friend, this study shows that drinking coffee stunts your growth. Explain what is wrong with your friend's interpretation.

9. If a person's arm span was *always* 3 inches shorter than his or her height, what would be the correlation between arm span and height? Explain.

10. Explain what is wrong with the following statements about correlation and regression.

   (a) The regression equation that describes the relationship between the age of a used Ford Mustang and its value is $y = -1000x + 10,000$, where the input is the age of the automobile (in years) and the output is its value (in dollars). The correlation describing this relationship is 0.83.

   (b) The correlation between the number of years of education and yearly income is 1.23.

11. Give an example of two variables not mentioned in the reading that have a negative correlation.

12. The following table represents the height and length of arm span (both in centimeters) for 16 college students. (This is the complete list from the Displaying Data Section.) Enter the data in your calculator and answer the following questions.

Height	152	156	165	163	165	168	168	173
Arm Span	159	155	160	166	163	176	164	171
Height	173	173	173	179	180	182	183	193
Arm Span	170	169	176	183	175	181	188	188

(a) What is the correlation? What does this number tell you about the relationship between a person's height and the length of his or her arm span?

(b) Determine the linear regression equation for this data with height as your input (or independent variable) and arm span as the output (or dependent variable).

13. Input the following data into your calculator.

x	1	2	3	4	5	6	7	8
y	5	7	9	11	13	15	17	19

(a) Determine the regression equation and correlation for the data.

(b) What does the correlation tell you about the data and the regression equation?

14. The regression line comparing the number of assignments not turned in and the final grade in an introductory college math class was:

$$y = -0.248x + 3.01$$

The correlation is -0.473.

(a) The correlation is negative. What does this mean?

(b) What is the meaning (in terms of number of assignments and final grade) of the 3.01?

(c) What is the meaning (in terms of number of assignments and final grade) of the -0.248?

(d) According to the regression line, what is the final grade of a student who doesn't turn in three assignments?

(e) How well do you think this line fits the data? Justify your answer.

15. In the reading, we found that the least-squares regression line for the carbon emissions data shown in the following table was $y = 7.25x + 207$, where x was the number of years since 1992 and y was the amount of carbon emissions for that year.

Year	1992	1993	1994	1995	1996	1997	1998	1999	2000	2001
Carbon Emissions (Million Metric Tons)	211	217	220	225	233	245	250	252	264	280

(a) The correlation for the regression line was $r = 0.98$. What information does this give you about the carbon emissions data?

(b) Use the regression line to estimate the level of carbon emissions in the commercial sector in 2008. How confident are you about the accuracy of your answer? Explain.

Regression and Correlation: Activities and Class Exercises

1. **Olympic 100-Meter Run.** The following table represents the winning times (in seconds) in the 100-Meter Run in the Olympic Games from 1928 to 2000.[4]

Year	Men's Winning Time	Women's Winning Time
1928	10.8	12.2
1932	10.3	11.9
1936	10.3	11.5
1948	10.3	11.9
1952	10.4	11.5
1956	10.5	11.5
1960	10.2	11.0
1964	10.0	11.4
1968	9.95	11.0
1972	10.14	11.07
1976	10.06	11.08
1980	10.25	11.6
1984	9.99	10.97
1988	9.92	10.54
1992	9.96	10.82
1996	9.84	10.94
2000	9.87	10.75

(a) Determine the regression equation where the number of years since 1900 is the input and the men's winning time is the output. Do the same thing for the women's winning times.

---

[4] *100M*, 15 July 2002, <http://www.kented .org.uk/ngfl/numeracy/sheets/olympics.xls>.

(b) Why does it make sense that the slopes of your regression lines are negative?

(c) Using each of your regression equations, determine the predicted winning times for the men's and women's 100-meter run in the 1952 Olympic Games. Are these numbers close to the actual winning times?

(d) Suppose the Olympic Games were held in the year 1000 A.D.

    i. What input represents 1000 A.D.?

    ii. Using each of your regression equations, determine the predicted winning times for the men's and women's 100-meter run.

    iii. Do these numbers seem reasonable for winning times? Explain.

(e) We want to predict the winning time for 2100.

    i. What input represents 2100?

    ii. Using each of your regression equations, determine the predicted winning times for the men's and women's 100-meter run for the Olympic Games in the year 2100.

    iii. Do these numbers seem reasonable for winning times? Explain.

(f) Based on your answers to the previous questions, when do you think regression equations work well in predicting outcomes?

2. **Measurement.** Throughout history, different measures of length (such as the foot) have been linked to human anatomy. Leonardo da Vinci (1452–1519) was quite interested in the proportions of the human body. He wrote about the views of the Roman architect and engineer Vitruvius Pollio (1st Century B.C.):

> Vitruvius declares that Nature has thus arranged the measurements of a man: four fingers make one palm; four palms make one foot; six palms make one cubit; four cubits make once a man's height; four cubits make a pace, and twenty-four palms make a man's height.[5]

If Vitruvius is correct that four palms make one foot and twenty-four palms make one's height, then six times the length of a person's foot should be about the same as his or her height. To see if this statement is true, do the following:

(a) Get measurements of foot length and height (both in centimeters) from a number of people. Does six times the length of a person's foot appear to be approximately equal to his or her height?

(b) Make a scatterplot with foot length on the horizontal axis and height on the vertical axis.

(c) Given the statement, "Six feet make one's height," answer the following questions.

    i. What linear equation represents this statement?

    ii. Graph this equation with your scatterplot. Does the line seem to fit the data? Does this statement seem accurate?

---

5 Klein, H. Arthur, *The World of Measurements,* New York: Simon and Schuster, 1974. 68.

iii. Find the regression equation for your data. Describe what this equation means in terms of foot length and height.

iv. How well does your data fit your regresion line? Explain.

3. **Running the Mile.** The following table lists years and speed for running the mile. The speeds, given in miles per hour, are the world-record for that year. There is also a scatterplot of these data. The years on the scatterplot are given as number of years since 1800 (i.e., $t = 0$ corresponds to 1800, $t = 100$ corresponds to 1900, etc.).[6]

Years	Running Speed mph		Years	Running Speed mph
1865	13.02		1943	14.84
1868	13.39		1944	14.90
1874	13.53		1945	14.91
1875	13.61		1954	15.13
1880	13.68		1957	15.18
1882	13.77		1958	15.35
1884	13.93		1962	15.36
1894	13.94		1964	15.38
1895	14.08		1965	15.41
1911	14.10		1966	15.56
1913	14.15		1967	15.58
1915	14.25		1975	15.69
1923	14.38		1979	15.72
1931	14.50		1980	15.73
1933	14.54		1981	15.84
1934	14.59		1985	15.91
1937	14.61		1993	16.04
1942	14.62		1999	16.13

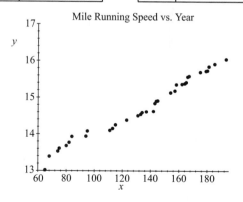

Mile Running Speed vs. Year

6 *World Records for ONE Mile Race*, 18 July 2002, <http://www.ex.ac.uk/cimt/data/worldrec/mmile.htm>.

(a) Find the regression line that fits this data. Your input should be number of years since 1800 (i.e., $t = 0$ corresponds to 1800, $t = 100$ corresponds to 1900, etc.) and your output should be speed in miles per hour.

   i. What is the physical meaning of the slope of the regression line?

   ii. What is the physical meaning of the y-intercept of the regression line?

   iii. Is the regression line a good fit for your data? Justify your answer.

(b) World records for the mile are typically given as time rather than as speed.

   i. What was the fastest time (in minutes) for running the mile in 1865?

   ii. What was the fastest time (in minutes) for running the mile in 1993?

   iii. The regression line you found in part(a) gave the speed for that year. How would you convert this equation so that the input was time (since 1800) and the output was time (in minutes)?

(c) When Roger Bannister broke the 4-minute mile, the news made headlines around the world. What year did this occur?

4. **Smoking Trends.** The following table is the prevalence of cigarette smoking by persons 25 years old and older in the United State from selected years 1974 to 1999.[7]

Year	'74	'79	'83	'85	'90	'92	'94	'95	'97	'98	'99
Percent of Americans who smoke	36.9	33.1	31.6	30	25.4	26.3	24.9	24.5	24	23.4	22.7

(a) Find the regression equation where the number of years since 1970 is the input and the percent of Americans who smoke is the output.

(b) Why do you think the information was given in percent of Americans who smoke rather than number of smokers?

(c) What is the correlation? What does this tell you about your data?
The following data table is from the same study.

Year	'74	'79	'83	'85	'90	'92	'94	'95	'97	'98	'99
Percent of Americans with a H.S. diploma or GED who smoke	36.2	33.6	33.5	32.0	29.1	30.5	29.1	29.1	29.9	28.9	28.0
Percent of Americans with a bachelor's degree or higher who smoke	27.2	22.6	20.5	18.5	13.9	15.2	11.9	13.6	11.4	10.9	11.1

(d) Find the regression equation for each row of data where the input is the number of years since 1970 and the output is the percent of people who smoke.

---

[7] *Table 61. Age-adjusted prevalence of current cigarette smoking by persons 25 years of age and over, according to sex, race, and education: United States, selected years 1974–2001*, 15 March 2004, <http://www.cdc.gov/nchs/data/hus/tables/2003/03hus060.pdf>.

      i. What does the regression line indicate about level of education and the likelihood of smoking?

     ii. In 1999, there were 29,538,180 people who had a bachelors degree and 57,860,082 people who had earned a high school diploma or GED.[8]

        A. Calculate the number of smokers with a bachelor's degree and number of smokers with a high school diploma or GED.

        B. Does your answer for part (A) contradict your statement in part (i)? Why or why not?

    iii. Draw the scatterplot and regression line for a bachelor's degree (input is still years and output is still percent of smokers).

    iv. Looking at the graph and plots drawn in (iii), do you think the slope of a regression line just for the data from years 1990 to 1999 would be different from the slope of the regression line for data from 1974 to 1999? Explain.

     v. Find the regression line for years 1990 to 1999 for a bachelor's degree. What does the slope tell you about recent trends in smoking?

5. **Destruction of the Ozone Layer.** The following table gives the maximum area of the ozone hole over Antarctica. The area is in million square kilometers.[9]

Year	Area (Millions sq. km)	Year	Area (Millions sq. km)
1980	2.954	1991	23.184
1981	3.049	1992	25.506
1982	9.950	1993	25.913
1983	11.709	1994	25.266
1984	14.546	1995	23.481
1985	19.009	1996	28.170
1986	14.984	1997	26.168
1987	22.876	1998	29.361
1988	14.214	1999	26.670
1989	22.385	2000	30.924
1990	21.325	2001	28.020

   (a) Let time be measured in years since 1980. So 1980 is represented by $t = 0$, 1990 by $t = 10$, etc. Find the equation for the regression line where years since 1980 is the input and the area of the ozone hole is the output.

---

[8] US Census Bureau. *Statistical Abstract of the United States.* Washington, DC: Hoover's Business Press, 2000.

[9] *Ozone Indicators – Size of the Antarctic ozone hole*, 15 March 2004,
<http://www.environment.govt.nz/indicators/ozone/hole-table.html>

(b) Using your equation from part (a), compute the area of the ozone hole for 1980, 1987, 1994, and 2001. For each year, find the difference between your result and the actual area of the hole given in the table. What does this tell you about your regression line?

(c) What is the correlation for your regression line? What information does this give you about how well the line fits the data?

(d) Using your equation from part (a), predict the area of the hole in the year 2015. How confident are you about the accuracy of your answer? Why?

6. **Enrollment.** The table below shows the enrollment at Michigan State University for the years between 1994 and 2001.[10]

Year	1994	1995	1996	1997	1998	1999	2000	2001
Enrollment	40,254	40,647	41,545	42,603	43,189	43,038	43,366	44,223

(a) Let time be measured in years since 1990. So 1994 is represented by $t = 4$, 2000 by $t = 10$, etc. Find the equation for the least squares regression line where years since 1990 is the input and enrollment is the output.

(b) What is the correlation? Do these data appear to be linear? Why or why not?

(c) What is the physical meaning of the slope?

(d) Use your regression equation to predict the number of students enrolled at Michigan State University in 1990. How confident are you about the accuracy of your answer? Explain.

(e) Use your regression line to predict the number of students enrolled at Michigan State University in 2030. How confident are you about the accuracy of your answer? Explain.

---

[10] *MSU—Office of the Registrar*, 15 March 2004, <http://www.reg.msu.edu/Reports/ROReports/genreport.asp>.

9

Exponential Functions

Money in a savings account, the cost of college tuition, the size of a population, and the decay of a radioactive substance are all examples of situations which typically increase (or decrease) by a fixed percentage. For example, the interest earned is equal to a percentage of the money in the account. Next year's tuition increase is a percentage of this year's tuition. Functions that exhibit this type of behavior, increasing (or decreasing) by a fixed percentage, are called exponential functions. Exponential functions have the property of eventually increasing (or decreasing) very rapidly, something that is fortunate if you have money in the bank or distressing if you consider the repercussions of unfettered population growth. In this section, we will examine properties of exponential functions and show how to determine if a given function is exponential.

Definition of Exponential Functions

A New Look at an Old Story

Our variation on an old story involves the owner of a baseball team who, with ten games left in the regular season, tries to sign a young baseball player to a contract. The player convinces the team owner to pay him in an unorthodox manner. Rather than an annual salary, the player asks the owner to give him one dollar for signing his contract and to double the amount he receives per game each time the team plays. The owner, seeing a chance to secure a reasonably good player for what seems like an insignificant wage, readily agrees. The owner thinks "Given that the player will only receive $2 for the first game and there are only ten games left in the regular season, how expensive could this be?" Table 9.1 shows how much the player received for each of the ten games. Remember that each game's payment is double the previous payment. The payment for Game 0 represents the $1 bonus he received for signing the contract.

Altogether, the player received $1 + $2 + $4 + $8 + \cdots + $1024 = 2047 for playing ten games. Notice that for playing the 10th game, the player received $1024—quite an

TABLE 9.1

The baseball player receives a $1 bonus for signing and his payment doubles each time he plays a game.

Game	0	1	2	3	4	5	6	7	8	9	10
Payment	$1	$2	$4	$8	$16	$32	$64	$128	$256	$512	$1024

improvement over the $2 he received for the first game! However, the real impact of this method of payment is seen when there are additional games. To understand this, look at the pattern that was generated. Let P be the function where $P(n)$ is the payment the player received for the nth game. Then $P(0) = 1$ since he received one dollar for signing the contact. To find $P(1)$, we double the amount he received for signing. So $P(1) = 2P(0) = 2 \cdot 1 = 2$. Following this pattern, we have:

$$P(2) = 2P(1) = 2 \cdot 2 = 2^2 = 4.$$

$$P(3) = 2P(2) = 2 \cdot 2^2 = 2^3 = 8.$$

$$P(4) = 2P(3) = 2 \cdot 2^3 = 2^4 = 16.$$

Notice the pattern. The pay for game 4, $P(4)$, is 2^4, the pay for game 5, $P(5)$, is $2 \cdot 2^4 = 2^5$, etc. In general, the nth game's pay is:

$$P(n) = 2P(n-1) = 2^n.$$

This method of payment is not too bad for the owner when the player only plays ten games. However, the story does not end here. With the young player's help, the team made the playoffs and went on to win the World Series, playing a total of nineteen additional games. The owner quit bragging to his friends about how he got a star player for such cheap wages when his accountant pointed out to him that he would owe the young player $2^{29} = \$536,870,912$ for just playing the final game. In fact, the player earned a total of $1,073,741,823 for the 10 regular games and 19 playoff games, becoming the highest paid sports player of all time. He then retired as a baseball player, bought the team from the bankrupt owner, invested the remainder of his money in stocks and bonds (which grow exponentially), and enjoyed being a fiscally responsible owner of a baseball team.

The baseball player's payment function, $P(n) = 2^n$, is an example of an exponential function. Exponential functions always have a constant growth factor. The **growth factor** is the ratio, $f(x+1)/f(x)$. A constant growth factor simply means that, for every x, $f(x+1)$ is a fixed multiple of $f(x)$. In our baseball example, the growth factor was 2 since the next game's salary was 2 times the previous game's salary.

Comparing Linear and Exponential Functions

Exponential functions are similar to linear functions in that they both have a constant rate. However, these rates measure different things, which leads to different types of behavior.

TABLE 9.2

Linear function where $L(0) = 1$ and the rate of change is two.

x	0	1	2	3	4	5	6	7	8	9	10
$L(x)$	1	3	5	7	9	11	13	15	17	19	21

TABLE 9.3

Exponential function where $E(0) = 1$ and the growth factor is 2.

x	0	1	2	3	4	5	6	7	8	9	10
$E(x)$	1	2	4	8	16	32	64	128	256	512	1024

Recall that a linear function has a constant *rate of change*,

$$\frac{y_2 - y_1}{x_2 - x_1} = \frac{f(x_2) - f(x_1)}{x_2 - x_1}.$$

This means that, every time the input increases by one, a fixed amount will be *added* to the previous output. For example, let L be the linear function where $L(0) = 1$ and the rate of change, or slope, is 2 (i.e., the change in the output is twice the change in the input). Some values of $L(x)$ are given in Table 9.2.

An exponential function, on the other hand, has a constant *growth factor*, $f(x+1)/f(x)$. This means that, every time 1 is added to the input, the previous output will be *multiplied* by a fixed amount. Let E denote the exponential function for which $E(0) = 1$ and whose growth factor is 2 (i.e., $E(x + 1)$ is double $E(x)$). Some values of $E(x)$ are given in Table 9.3. It does not take long to see that multiplication by a number bigger than 1 produces much larger outputs than addition!

Graphs of the two functions, $y = E(x)$ and $y = L(x)$, are shown in Figure 9.1. Even though the linear function starts out larger for small values of x, the exponential

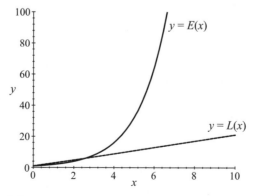

FIGURE 9.1

A graph of the exponential function E, and the linear function, L.

function soon becomes much larger—since it is concave up. A constant growth factor is significantly more powerful than a constant rate of change if you are interested in functions that increase quickly. In fact, any exponential function (with a growth factor larger than one) will eventually become larger than any linear function. Thankfully, exponential functions are used to model money accumulated through interest. Unfortunately, exponential functions are also used to model accumulated debt.

We saw in the baseball example that $P(n) = 2^n$ was the exponential function that started with \$1 and had a growth factor of 2. Let's generalize this by finding the formula for the exponential function, $y = E(n)$, where $E(0) = b$ and $y = E(n)$ has a growth factor of a. For now, we will assume that $a > 0$ and $a \neq 1$. Our starting value is $E(0)$, which was given to us as b. Since the growth factor is a, the next output has to be a times the previous output. That means:

$$E(1) = a \cdot E(0) = a \cdot b.$$

Similarly,

$$E(2) = a \cdot E(1) = a \cdot ab = a^2 b,$$
$$E(3) = a \cdot E(2) = a \cdot a^2 b = a^3 b.$$

In general,

$$E(n) = a^n b = ba^n.$$

Note that $a > 1$ produces growth (i.e., the values of E increase) while $0 < a < 1$ produces decay (i.e., the values decrease). Although we have derived the formula for $y = E(n)$ using only integers, the same formula works for all real numbers. Values for $y = E(n)$ are summarized in Table 9.4.

TABLE 9.4
Exponential function where $E(0) = b$ and the growth factor is a.

x	0	1	2	3	4	5	6	$\cdots$	n
$E(x)$	b	ab	$a^2 b$	$a^3 b$	$a^4 b$	$a^5 b$	$a^6 b$	$\cdots$	ba^n

We are now ready to give the formal definition of an exponential function. An **exponential function** is a function, $y = f(x)$, for which the growth factor, $f(x+1)/f(x) = a$, is constant. Symbolically, an **exponential function** is

$$f(x) = ba^x$$

where $a > 0$ and $a \neq 1$. The y-intercept (or starting value) is b and the growth factor is a. So $b = f(0)$ and $a = f(x+1)/f(x)$. The growth factor, a, is also called the **base** of the exponential function. A growth factor greater than 1 gives *exponential growth* while a growth factor between 0 and 1 gives *exponential decay*.

Properties of Exponential Functions

Now that we have the definition and a general formula for an exponential function, we are interested in exploring its properties. For the moment, we will assume that the starting amount, b, is equal to 1. In other words, we will start by only looking at functions of the form $E(x) = a^x$.

Growth Factors Greater Than 1

Let's compare the three exponential functions $f(x) = 2^x$, $g(x) = 3^x$ and $h(x) = 4^x$. Some values for these functions are given in Table 9.5. Graphs of $y = f(x)$, $y = g(x)$, and $y = h(x)$ are shown in Figure 9.2.

TABLE 9.5
Some values for $f(x) = 2^x$, $g(x) = 3^x$ and $h(x) = 4^x$.

x	0	1	2	3	4
$f(x) = 2^x$	1	2	4	8	16
$g(x) = 3^x$	1	3	9	27	81
$h(x) = 4^x$	1	4	16	64	256

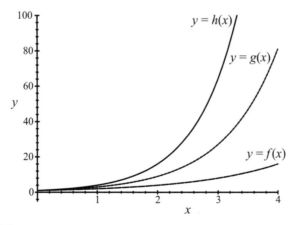

FIGURE 9.2
Graphs of $f(x) = 2^x$, $g(x) = 3^x$, and $h(x) = 4^x$ on a domain of $0 \leq x \leq 4$.

It is obvious that an exponential function with a larger growth factor will grow faster. You can see this from Table 9.5—the larger the growth factor, the larger the outputs.

Also notice from Table 9.5 that all three functions contain the points $(0, 1)$ (the first column) and $(1, a)$ (the second column). This (as well as the other properties of exponential functions) results from the properties of exponents. Looking at this symbolically, if $E(x) = a^x$, then $E(0) = a^0 = 1$ and $E(1) = a^1 = a$. These properties should also make sense when you remember that an exponential function of the form $E(x) = a^x$ has a starting value of 1 and a growth rate of a.

You should also observe that all of the outputs are positive. Since $a > 1$, $E(x) = a^x$ will always be positive. In other words, starting with one unit and multiplying by a positive growth factor guarantees that your output will be positive.

All three of these functions are increasing and concave up. The graphs are increasing since, for a growth factor greater than 1, a larger input produces a larger output. Think of this as "the more days something has been growing, the more you have." The graphs are concave up because the rate of change is increasing. You can see this by observing that the differences between successive outputs are increasing in Table 9.5. If you multiply a large amount by a growth factor of a, the increase is greater than if you multiply a small amount by a growth factor of a.

So far, we have only looked at positive inputs. However, exponential functions are also defined for negative values of x. Table 9.6 shows the same three functions, but this time with negative inputs. Approximations are rounded to three decimal places. A graph of $y = f(x)$, $y = g(x)$, and $y = h(x)$ using negative inputs is shown in Figure 9.3.

Notice we have the opposite behavior when using negative inputs. This time, the larger the growth factor, the *smaller* the output. Why is this happening? Recall that raising a

TABLE 9.6
Some values for $f(x) = 2^x$, $g(x) = 3^x$ and $h(x) = 4^x$ for negative inputs.

x	0	-1	-2	-3	-4
$f(x) = 2^x$	1	$\frac{1}{2} = 0.500$	$\frac{1}{4} = 0.250$	$\frac{1}{8} = 0.125$	$\frac{1}{16} \approx 0.063$
$g(x) = 3^x$	1	$\frac{1}{3} \approx 0.333$	$\frac{1}{9} \approx 0.111$	$\frac{1}{27} \approx 0.037$	$\frac{1}{81} \approx 0.012$
$h(x) = 4^x$	1	$\frac{1}{4} = 0.250$	$\frac{1}{16} \approx 0.063$	$\frac{1}{64} \approx 0.016$	$\frac{1}{256} = 0.004$

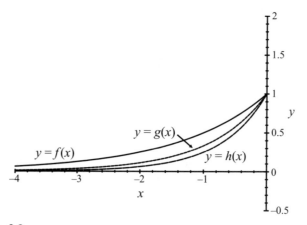

FIGURE 9.3
Graphs of $f(x) = 2^x$, $g(x) = 3^x$, $h(x) = 4^x$ on a domain of $-4 \leq x \leq 0$.

number to a negative power means you raise the *reciprocal* of the number to the positive power. In other words, $2^{-1} = \frac{1}{2}$, $2^{-2} = \left(\frac{1}{2}\right)^2 = \frac{1}{4}$, and $2^{-3} = \left(\frac{1}{2}\right)^3 = \frac{1}{8}$. For negative inputs, the functions $y = 2^x$, $y = 3^x$, and $y = 4^x$ are giving you powers of $\frac{1}{2}$, $\frac{1}{3}$, and $\frac{1}{4}$. Larger growth factors now give larger denominators causing the reciprocal to be actually smaller. Therefore, the larger the growth factor, a, the smaller the output for negative values of x. Notice from both Table 9.6 and Figure 9.3 that the outputs continue to be positive since we are still multiplying positive numbers.

Notice that all three functions get closer to the x-axis as x approaches negative infinity (i.e., x moves to the left). These functions all have the x-axis as a horizontal asymptote. In general, the graph of a function has a **horizontal asymptote**, $y = c$, if the values of the function get closer and closer to c as x approaches either positive infinity or negative infinity. For example, $f(x) = 2^x$ has the x-axis ($y = 0$) as a horizontal asymptote since, as x approaches negative infinity, $y = f(x)$ gets closer and closer to zero.

Growth Factors Less Than 1 (and Greater Than 0)

So far we have examined functions where a is greater than 1. Look at a table of data (Table 9.7) and graphs (Figure 9.4) of the functions $f(x) = \left(\frac{1}{2}\right)^x$, $g(x) = \left(\frac{1}{3}\right)^x$, and $h(x) = \left(\frac{1}{4}\right)^x$. Each of these functions has a growth factor that is less than 1 but greater than 0. Since our growth factor is now smaller than 1, the next output will be *smaller* than the previous output. In other words, the output is decaying rather than growing.

TABLE 9.7

Values for $f(x) = \left(\frac{1}{2}\right)^x$, $g(x) = \left(\frac{1}{3}\right)^x$, and $h(x) = \left(\frac{1}{4}\right)^x$ when $-4 \le x \le 4$.

x	-4	-3	-2	-1	0	1	2	3	4
$f(x) = \left(\frac{1}{2}\right)^x$	16	8	4	2	1	$\frac{1}{2} = 0.500$	$\frac{1}{4} = 0.250$	$\frac{1}{8} = 0.125$	$\frac{1}{16} \approx 0.063$
$g(x) = \left(\frac{1}{3}\right)^x$	81	27	9	3	1	$\frac{1}{3} \approx 0.333$	$\frac{1}{9} \approx 0.111$	$\frac{1}{27} \approx 0.037$	$\frac{1}{81} \approx 0.012$
$h(x) = \left(\frac{1}{4}\right)^x$	256	64	16	4	1	$\frac{1}{4} = 0.250$	$\frac{1}{16} \approx 0.063$	$\frac{1}{64} \approx 0.016$	$\frac{1}{256} = 0.004$

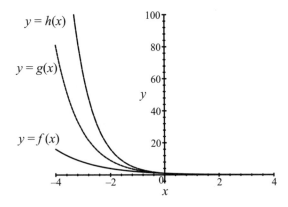

FIGURE 9.4

Graphs of $f(x) = \left(\frac{1}{2}\right)^x$, $g(x) = \left(\frac{1}{3}\right)^x$, and $h(x) = \left(\frac{1}{4}\right)^x$ on a domain of $-4 \le x \le 4$.

In comparing these functions, we again see that a larger growth factor gives a larger output when $x > 0$ and a smaller output when $x < 0$. Just as when the growth factors were greater than 1, these functions still contain the points $(0, 1)$ and $(1, a)$ where a is the growth factor. The outputs are still always positive. These properties are the same because the properties of exponents are not affected by the size of the growth factor.

These functions are different, however, in that they are decreasing. If you raise a number between 0 and 1 to a larger exponent, you get a smaller value. Your amount is decaying rather than growing. For example, suppose the baseball player had agreed to a salary of $10,000 for the first game and a growth factor of $\frac{1}{2}$. For the second game, he would have received $5,000, for the third game he would receive $2,500, and so on. By the tenth game, his payment would only be $19.53 and by the time he finished the World Series, he would be playing for less than a penny!

These new functions are still concave up. This means that the rate of change is increasing. Since the rate of change is increasing and the function is decreasing, the larger the input, the smaller the decrease. The functions still have a horizontal asymptote at $y = 0$, but this time it occurs as x approaches positive infinity (i.e., x moves to the right). The graphs of $y = (\frac{1}{2})^x$, $y = (\frac{1}{3})^x$, and $y = (\frac{1}{4})^x$ are similar to the graphs of $y = 2^x$, $y = 3^x$, and $y = 4^x$ except they have been "flipped" over the y-axis.

Let's summarize the properties of $f(x) = a^x$ where $a > 0$ and $a \neq 1$. Even though we only looked at a few examples to deduce these properties, they are true for all exponential functions. In particular, these properties are true even when a is not an integer but is any positive real number.[1]

- Comparing exponential functions, $f(x) = a^x$ and $g(x) = c^x$, where $a > c > 0$:
  * $a^x > c^x$ if $x > 0$. The larger the growth factor, the larger the output when your input is positive.
  * $a^x < c^x$ if $x < 0$. The larger the growth factor, the smaller the output when your input is negative.
- Points in common
  * $(0, 1)$
  * $(1, a)$
- Domain and Range
  * The domain is all real numbers.
  * The range is $f(x) > 0$.
- Increasing/Decreasing
  * $f(x) = a^x$ is increasing for all x when the growth factor, a, is greater than 1.
  * $f(x) = a^x$ is decreasing for all x when the growth factor, a, is between 0 and 1.
- Concavity
  * $f(x) = a^x$ is always concave up.

---

[1] This is because all of these properties are consequences of the properties of exponents.

The Significance of *b*

We looked at exponential functions of the form $f(x) = a^x$ for various values of a. Now let's return to the more general exponential functions of the form $f(x) = ba^x$ and explore the significance of b. For simplicity, we will restrict b to be positive. As before, we will start with a table, this time using the functions $f(x) = \frac{1}{3} \cdot 2^x$, $g(x) = 2^x$, and $h(x) = 3 \cdot 2^x$. (See Table 9.8.) Approximations are rounded to three decimal places. The graphs of these three functions for $-4 \le x \le 4$ are shown in Figure 9.5.

TABLE 9.8

Some values for $f(x) = \frac{1}{3} \cdot 2^x$, $g(x) = 2^x$, and $h(x) = 3 \cdot 2^x$.

x	$f(x) = \frac{1}{3} \cdot 2^x$	$g(x) = 2^x$	$h(x) = 3 \cdot 2^x$
-4	$\frac{1}{48} \approx 0.021$	$\frac{1}{16} \approx 0.063$	$\frac{3}{16} \approx 0.188$
-3	$\frac{1}{24} \approx 0.042$	$\frac{1}{8} = 0.125$	$\frac{3}{8} = 0.375$
-2	$\frac{1}{12} \approx 0.083$	$\frac{1}{4} = 0.25$	$\frac{3}{4} = 0.75$
-1	$\frac{1}{6} \approx 0.167$	$\frac{1}{2} = 0.5$	$\frac{3}{2} = 1.5$
0	$\frac{1}{3} \approx 0.333$	1	3
1	$\frac{2}{3} \approx 0.667$	2	6
2	$\frac{4}{3} \approx 1.333$	4	12
3	$\frac{8}{3} \approx 2.667$	8	24
4	$\frac{16}{3} \approx 5.333$	16	48

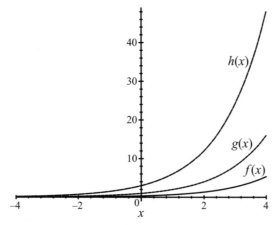

FIGURE 9.5

Graphs of $f(x) = \frac{1}{3} \cdot 2^x$, $g(x) = 2^x$, and $h(x) = 3 \cdot 2^x$.

Notice that the shape of the graph has not changed. Basically, all three graphs are similar to the shape of an exponential function whose growth factor is greater than 1. The most noticeable change is that the functions do not all cross the y-axis at the point $(0, 1)$. Rather, the functions go through the point $(0, b)$ since b is the starting point or y-intercept. Symbolically, if $f(x) = ba^x$, then $f(0) = ba^0 = b \cdot 1 = b$. Changing the y-intercept will not change the growth factor, but will change the actual outputs.

It is also no longer true that each function contains the point $(1, a)$. Instead, the functions now contain the point $(1, ba)$. You can easily find the value of both b and a from an appropriate graph or table of data. Recall that b is the y-intercept (or $f(0) = b$) and a is the growth factor, (or $a = f(x + 1)/f(x)$). Since the growth factor is constant, you can use any two points which are one unit apart to find a. Using data from Table 9.8, the ratio of $h(4)/h(3) = 48/24 = 2$ so the growth factor for $y = h(x)$ is 2.

How Do You Determine if Your Function is Exponential?

Suppose you are given a function either symbolically, numerically, or verbally. How do you determine if it's an exponential function?

Symbolic Representations

Let's start with a symbolic representation. In this case, make sure you have the representation in the form "$y =$" and see if it can be written as ba^x for some values of a and b.

Example 1.   Determine if $3^x y = 2$ is an exponential function. If so, give the growth factor and the y-intercept (i.e., the values of a and b). If not, briefly explain why.

Solution.   Solving $3^x y = 2$ for y, we get

$$ y = \frac{2}{3^x} = 2 \left(\frac{1}{3} \right)^x . $$

This is a decaying exponential function since it is in the form ba^x where the growth factor is a positive number less than 1. ∎

Example 2.   Determine if $x^3 y = 2$ is an exponential function. If so, give the growth factor and the y-intercept (i.e., the values of a and b). If not, briefly explain why.

Solution.   Solving $x^3 y = 2$ for y, we get

$$ y = \frac{2}{x^3} = 2 \left(\frac{1}{x} \right)^3 . $$

This is not an exponential function since it is not in the form ba^x. All exponential functions have x in the exponent and this one does not. ∎

Numerical Representations

Now let's look at functions represented numerically. The key to determining if you have an exponential function here is to remember that exponential functions have a constant growth factor. So you need to determine the ratios $f(x + 1)/f(x)$ to see if they are all the same. If so, your function is exponential and this constant is your value of a. To find b, find the output for $x = 0$. Remember, b is your y-intercept. This process is illustrated in the next example.

Example 3.   Using Table 9.9, determine if each function is an exponential function. If it is, find its equation. If not, briefly justify why.

TABLE 9.9

x	0	1	2	3	4
$f(x)$	8	1.6	0.32	0.064	0.0128
$g(x)$	8	6	4	2	0

Solution.   We need to see if the growth factor is constant. First, looking at $y = f(x)$ we see

$$\frac{f(1)}{f(0)} = \frac{1.6}{8} = 0.2, \quad \frac{f(2)}{f(1)} = \frac{0.32}{1.6} = 0.2,$$

$$\frac{f(3)}{f(2)} = \frac{0.064}{0.32} = 0.2, \quad \text{and} \quad \frac{f(4)}{f(3)} = \frac{0.0128}{0.064} = 0.2.$$

Since the ratio $f(x+1)/f(x)$ is constant, this is an exponential function with $a = 0.2$. To find the value of b, notice that $f(0) = 8$. So this is the exponential function $f(x) = 8 \cdot (0.2)^x$.

Now looking at $y = g(x)$ we see

$$\frac{g(1)}{g(0)} = \frac{6}{8} = 0.75 \quad \text{and} \quad \frac{g(2)}{g(1)} = \frac{4}{6} \approx 0.67.$$

Since the ratio $g(x+1)/g(x)$ is not constant, this is not an exponential function. In fact, this is the linear function $g(x) = -2x + 8$. ∎

Verbal Representations

To decide if a verbal description of a situation portrays an exponential function, ask yourself if this situation has a constant growth factor. Many things grow exponentially— money, human populations, bacteria, etc. For example, your savings account may be earning 3% interest, the population of your town might be increasing by 5% each year, or the population of an endangered species might be decreasing by 10% each year. Each of these has a constant growth factor. Therefore, we can use exponential functions to describe each of these situations. However, a 3% increase, a 5% increase, or a 10% decrease are not *growth factors*, but are *growth rates*.

Suppose a population of a town consists of 1000 people and is increasing annually by 5%. This means that next year there will be 5% more people in the town than this year. This makes next year's population

$$1000 + 1000(0.05) = 1000 + 50 = 1050.$$

We can also think of this as

$$1000 + 1000(0.05) = 1000(1 + 0.05) = 1000(1.05) = 1050.$$

This second method shows that while 5% = 0.05 is the growth rate, 1.05 is the growth factor since the town's population next year will be 105% of what it was this year. A growth factor is always one more than its associated growth rate.

Example 4. Nationally, state universities increased tuition by an average of 6.3% per year from 1990 to 2000.[2] In 1990, the average yearly tuition at state universities in the country was approximately $2159. Let C be the function whose input is the number of years since 1990, t, and whose output is the estimated cost of tuition, $C(t)$. Is this an exponential function? If so, find its equation. If not, explain why.

Solution. This is an exponential function since the growth factor is constant. The 6.3% = 0.063 increase given to us, however, is not a growth factor but is a growth rate. The growth factor is therefore 1.063. This means the 1991 tuition was 106.3% as much as the tuition in 1990. With the same reasoning, you could say the tuition in 1992 was 106.3% as much as the tuition in 1991, and so on for each succeeding pair of years. The starting value, $C(0)$, is $2159 if we choose to use 1990 as our starting point. The exponential function that describes the yearly tuition is

$$C(t) = 2159(1.063)^t,$$

where t is the number of years after 1990, and $C(t)$ is the cost of the average yearly tuition (in dollars) at a state public university. ∎

Example 5. In 1980, there were approximately 1,200,000 African elephants. In 1990, there were approximately 700,000 African elephants.[3] Assuming that the population growth factor is constant, find the exponential function that models this data and use it to predict the number of African elephants in the year 2000.

Solution. Assuming that the population growth factor is constant means we are assuming an exponential function. Let P be the function whose input is time (in years since 1980) and whose output is the estimated number of elephants. We are given that $P(0) = 1,200,000$ so

$$P(t) = 1,200,000a^t.$$

We are not given the growth factor but, instead, we know that $P(10) = 700,000$. So when $t = 10$, we have

$$1,200,000a^{10} = 700,000.$$

---

[2] *Home, Digest of Education Statistics Tables and Figures*, 16 March 2004, <http://nces.ed.gov/programs/digest/>.
[3] Burton, John A. *The Atlas of Endangered Species*. Quarto Publishing, 1991, p. 138.

Dividing both sides of our equation by 1,200,000 and taking the tenth root of both sides (or raising both sides to the one-tenth power), we get

$$1,200,000a^{10} = 700,000$$

$$a^{10} = \frac{700,000}{1,200,000}$$

$$a = \sqrt[10]{\frac{700,000}{1,200,000}}$$

$$a = \left(\frac{700,000}{1,200,000}\right)^{\frac{1}{10}}$$

$$a \approx 0.95.$$

This means that each year, the elephant population is approximately 95% of what it was the previous year. Our function is

$$P(t) = 1,200,000 \cdot (0.95)^t.$$

The population in the year 2000 (assuming the growth factor remains constant) would be

$$P(20) = 1,200,000 \cdot 0.95^{20} \approx 430,000. \quad \blacksquare$$

Notice in the last example that we were never given the growth factor. As long as you know your function is exponential, knowing any two points on your function will allow you to algebraically solve for both b and a as illustrated in Example 5.

Example 6. An employee was offered the choice of receiving either a $1000 bonus added to his base pay every year or a 2.3% cost of living increase added to his base pay every year. He chose to take the $1000 bonus. His annual salary in 1997 was $33,500. Let S be the function whose input is the number of years since 1997 and whose output is the employee's base salary for that year. Is S an exponential function? If so, find it's equation. If not, explain why.

Solution. The employee's salary is being increased at a constant rate of $1000 per year. However, this amount is *added* to his salary. This means the 1000 is a constant rate of change rather than a growth factor. So S is a linear function, not an exponential function. The equation for S is

$$S(t) = 33,500 + 1000t$$

where t is the number of years since 1997. $\blacksquare$

Shifting Exponential Functions

In the section *Applications of Graphs*, we discussed what happens when you change either the input or the output of a function by a constant. Let's look at how this changes exponential functions. We will start with the exponential function $f(x) = 3^x$ whose graph is shown in Figure 9.6.

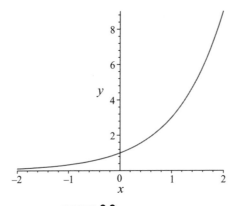

FIGURE 9.6
A graph of $f(x) = 3^x$.

Let's first see what happens when we add 2 to the output. The graph of $g(x) = 3^x + 2$ is shown in Figure 9.7(b). Notice that this raises the graph by 2 units. However, the shape of the graph remains the same. The most noticeable difference is that the horizontal asymptote as $x \to -\infty$ is now $y = 2$ rather than $y = 0$.

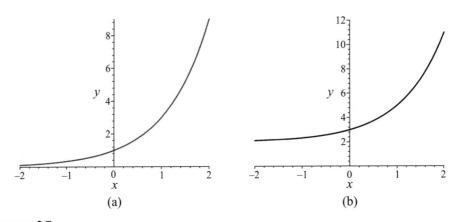

(a)   (b)

FIGURE 9.7
(a) A graph of $f(x) = 3^x$. (b) A graph of $g(x) = 3^x + 2$. Notice that the shape of the graph remains the same but the horizontal asymptote is at $y = 2$ rather than $y = 0$ for g.

If we add 2 to the input instead of the output, the resulting graph is shifted 2 units to the left (see Figure 9.8). Our new function is $g(x) = 3^{x+2}$. Using properties of exponents, we find that

$$g(x) = 3^{x+2} = 3^x \cdot 3^2 = 3^x \cdot 9 = 9 \cdot 3^x.$$

So this is just an exponential function with a different y-intercept. In other words, this is an exponential function of the form $y = b \cdot a^x$. The shape of the graph remains the same and we can consider this as a new exponential function (rather than a horizontal shift of our original function).

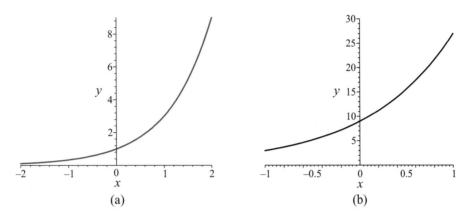

FIGURE 9.8

(a) A graph of $f(x) = 3^x$. (b) A graph of $g(x) = 3^{x+2}$. The second graph has the same shape as the original graph, but has a different y-intercept since it has been shifted to the left.

Next, let's multiply the output by 2. In this case, $g(x) = 2 \cdot 3^x$. Again, this is just an exponential function with a different y-intercept but the same growth factor (see Figure 9.9). The shape of the graph remains the same and we can again consider this as a new exponential function (rather than a vertical stretch of our original function).

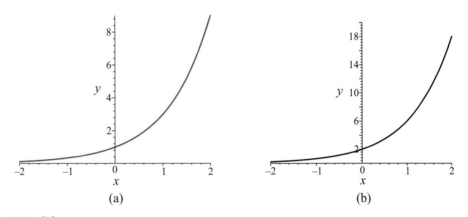

FIGURE 9.9

(a) A graph of $f(x) = 3^x$. (b) A graph of $g(x) = 2 \cdot 3^x$. The second graph has the same growth factor as the original graph, but has a different y-intercept.

Finally, let's see what happens when we multiply the input by a constant. Let $g(x) = 3^{2x}$. The graphs of $f(x) = 3^x$ and $g(x) = 3^{2x}$ are shown in Figure 9.10. Using properties of exponents, we find that $g(x) = 3^{2x} = (3^2)^x = 9^x$. So we can consider $g(x) = 3^{2x} = 9^x$ to be an exponential function with a different growth factor. The shape of the graph remains the same and we can consider this to be a new exponential function (rather than a horizontal compression of the original function).

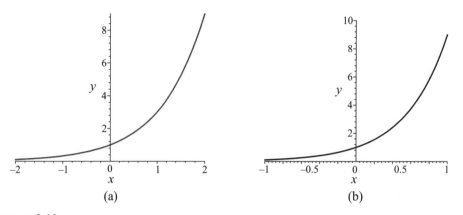

FIGURE 9.10
(a) A graph of $f(x) = 3^x$. (b) A graph of $g(x) = 3^{2x}$. The second graph has the same shape as the original graph, but has a different growth factor.

In summary, let $f(x) = b \cdot a^x$.

- Adding a constant, c, to the output, $g(x) = b \cdot a^x + c$, shifts the function vertically and changes the horizontal asymptote to $y = c$.
- Adding a constant, c, to the input, $g(x) = b \cdot a^{x+c} = b \cdot a^c \cdot a^x$, produces a new exponential function with a different y-intercept.
- Multiplying the output by a constant, $c > 0$, $g(x) = c(b \cdot a^x) = cb \cdot a^x$, produces a new exponential function with a different y-intercept.
- Multiplying the input by a constant, c, $g(x) = ba^{cx} = b \cdot (a^{cx}) = b(a^c)^x$, produces a new exponential function with a different growth factor.

Example 7.   Let $f(x) = 5^x$. If you multiply the input by 2, what is your new growth factor?

Solution.   Multiplying the input by 2 gives us $f(2x) = 5^{2x} = (5^2)^x = 25^x$. So your new growth factor is 25. ■

Example 8.   Find the equation of the form, $f(x) = ba^x + c$, given $f(0) = 4$, $f(1) = 13$, and the horizontal asymptote is $y = -1$.

Solution.   The horizontal asymptote gives the vertical shift. Therefore, we know that $c = -1$ and our function has the form

$$f(x) = ba^x - 1.$$

Now using the fact that $f(0) = 4$, we have

$$4 = ba^0 - 1$$
$$4 = b - 1$$
$$5 = b.$$

Finally, using the fact that $f(1) = 13$ we have

$$13 = 5a^1 - 1$$
$$14 = 5a$$
$$\frac{14}{5} = a.$$

This gives us $f(x) = 5(\frac{14}{5})^x - 1$. ∎

Application: Inflation

You have probably heard your grandparents saying something like, "In my day, a Big Mac cost a quarter!" But they also fail to tell you that a quarter was worth more than it is now. The concept of money changing in value through time is known as inflation. Table 9.10 contains data for every 5 years from 1900 to 2000. The dollar values for each year represent how much it would take, in buying power, to equal one dollar's value in 1900.[4]

TABLE **9.10**
The value of a 1900 dollar from 1900 to 2000.

Year	1900	1905	1910	1915	1920	1925	1930
Value	$1.00	$1.09	$1.25	$1.38	$2.60	$2.40	$2.16
Year	1935	1940	1945	1950	1955	1960	1965
Value	$1.85	$1.89	$2.44	$3.36	$3.60	$4.00	$4.27
Year	1970	1975	1980	1985	1990	1995	2000
Value	$5.34	$7.45	$11.59	$14.68	$17.96	$20.61	$23.36

Figure 9.11 shows a scatterplot of the data from Table 10 where the input is years after 1900 and whose output is the value of a 1900 dollar.

Just looking at the scatterplot, we can see that the data is roughly exponential in shape. Knowing this, we will develop an exponential function to model this data. Let V be the function whose input is time (in years since 1900) and whose output is the value of a dollar. We have two variables to find: the initial value and the growth factor. The initial value is $V(0)$ (the y-intercept), or in this case, the value of the dollar in the year 1900. This value is given to us in the table as simply $1, so $V(0) = 1$ and $V(t) = 1 \cdot a^t = a^t$. To find the growth factor, a, we can choose the value of a dollar in any given year. Each value we choose will give us a slightly different value for a, since our data does not perfectly fit an exponential function. We will use the value of a dollar in the year 2000,

---

[4] *Global Financial Data,* 16 March 2004, <http://www.globalfindata.com>.

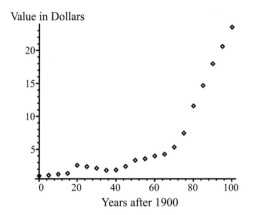

FIGURE **9.11**
Scatterplot showing how much it would take, in buying power, to equal one dollar's value in 1900.

or $f(100)$ (100 years after 1900). This gives us the equation, $23.36 = a^{100}$ since the buying power of a dollar in 2000 is $23.36. Taking the 100th root of both sides of our equation (or raising both sides to the one-hundredth power), we find

$$a = \sqrt[100]{23.36} = 23.36^{\frac{1}{100}} \approx 1.032.$$

This means that, each year, the dollar is worth approximately 1.032 times more than it was worth a year earlier. So our equation for the dollar value at time t is $V(t) = 1.032^t$. Figure 9.12 graphs this function along with the original scatterplot.

To determine an equation for the value of a dollar we could have also used exponential regression. This can be done on a calculator similar to how linear regression is done. Using the data from Figure 9.11 and performing exponential regression with a TI-83 calculator, we get the equation $y = 0.838 \cdot 1.03125^x$. This equation will not contain the first and last points from our data as does $V(t) = 1.032^t$, but does a better job of approximating many of the points between the first and last.

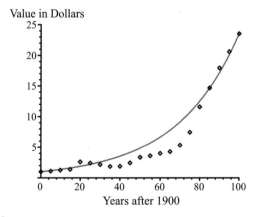

FIGURE **9.12**
The exponential function, $V(t) = 1.032^t$, along with our original scatterplot.

We want to use our function, $V(t) = 1.032^t$, to determine if things are more expensive merely because of inflation or if the price is actually higher in equivalent dollars. To do this, consider the equation $V(p, t) = p \cdot 1.032^t$, where p is the price of a certain product in 1900 and t is time since 1900.[5]

Example 9. The cost of a pound of coffee in 1900 was \$0.16 and had risen to \$3.33 per pound in 1997.[6] How well does the multivariable function, $V(p, t) = p \cdot 1.032^t$, where p is the price in 1900 and t is the time since 1900, work for estimating the cost of coffee in 1997?

Solution. We know p will be 0.16 and t will be 97, giving us $V(0.16, 97) = 0.16(1.032)^{97} \approx 3.40$. Thus, the estimated price for a pound of coffee in 1997 is \$3.40. This is very close to the actual cost of \$3.33 per pound. ∎

Example 10. The 2002 price for a scientific calculator is \$9.95. Using the modeling function $V(t, p) = p(1.032)^t$, what would the estimated price in 1972 have been?

Solution. The starting year for our multivariable function is no longer 1900, but rather 2002. Therefore, t represents years after 2002 if t is positive or before 2002 if t is negative, while p is the price of the product in 2002. Since 1972 is 30 years before 2002, we have $V(-30, 9.95) = 9.95(1.032)^{-30} = 3.87$, so our estimated price for a calculator in 1972 is \$3.87. ∎

Using this exponential model to predict prices too far into the future or past may not yield accurate results because the modeling function assumes a constant inflation rate and that no other factors influence price. This can lead to obviously incorrect answers such as the answer to Example 10. In 1972, scientific calculators cost nearly \$300 and today cost a mere \$9.95. Inflation does influence the price of technological devices but, in this case, this is not as important as research and the development of faster and cheaper products.

Reading Questions for Exponential Functions

1. If our baseball player participated in 10 regular season games and 14 playoff games before reaching the World Series, how much did he earn for playing the first game of the World Series?

2. How is a linear function similar to an exponential function? How is a linear function different from an exponential function?

3. Assume each of the following is an exponential function. In each case, give the growth factor (value of a) and the y-intercept (value of b).

   (a) $f(x) = 5^x$ (c) $f(0) = 2$, $f(1) = 6$

   (b) $f(x) = 3(\frac{1}{2})^x$ (d) $f(0) = 1$, $f(2) = 36$

---

[5] Note that t will be positive for years after 1900 and negative for years before 1900.

[6] Derks, Scott, Ed., *The Value of a Dollar: Prices and Incomes in the United States 1860–1999*, Grey House Publishing, Lakeville, CT, 1999 pp. 54, 461.

4. Let $f(x) = ba^t$ be an exponential function which models the population of an endangered species. What do you know about the value of a? Justify your answer.

5. Let $P(n) = 2^n$ represent the amount of money paid to our hypothetical baseball player who received \$1 for signing his contract and whose pay was doubled for every game the team played.

   (a) For what game did he first earn more than \$1,000,000?

   (b) We stated that he would earn \$536,870,912 for playing the last game of the World Series (the 29th game). What did he earn for playing the 27th game?

6. Let $f(x) = a^x$. The graph of f is given below. Is a greater than 1 or less than 1? How do you know?

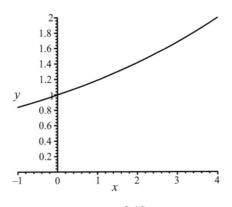

FIGURE 9.13

7. Give the equation for the exponential function whose y-intercept is 1.5 and whose growth factor is 4.

8. Let $f(x) = ba^x$ be the exponential function whose graph is given below. What is the value of b? How do you know?

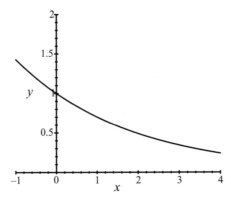

FIGURE 9.14

9. Determine if each of the following is an exponential function. If so, give its equation. If not, briefly explain why.

(a) $2y = 4^x$

(b)

x	-2	-1	0	1	2
$f(x)$	4	2	1	0.5	0.25

(c) The function describing the amount of money received by returning pop bottles if the deposit is $0.10 per bottle.

10. In 1997 DVD players were introduced to the electronic media market. In January of 1998 the average selling price of a DVD player was $600. By January of 2001, this price had decreased to $177. Below is a table showing the average price of a DVD player.[7]

Years	1998	1999	2000	2001	2002
Price	$600	$410	$295	$210	$177

(a) Assume the price of DVD players can be modeled by an exponential function, $p(t) = ba^t$, where t is years since 1998. Will a be less than one or greater than one? Explain.

(b) Find an exponential function that models the price of DVD players, $p(t) = ba^t$, where t is measured in years since 1998.

(c) Use the function you found in part(b) to predict the price of a DVD player in 2000. How well does this match the data in the table?

(d) Use your exponential function to predict the price in 2020. Does this seem reasonable? Explain.

11. Example 4 gave the growth rate for tuition at state universities. Assume you have a child who will be a freshman in college in the fall of 2027. How much will tuition cost if the growth rate remains the same?

12. Let $h(x) = 5(4)^x$. Find each of the following.

(a) $h(-3)$

(b) $h(-1)$

(c) $h(0)$

(d) $h(8)$

13. Determine if each of the following tables describes an exponential function. If so, give the equation for the function. If not, briefly explain why.

---

[7] "Taking Command." *U.S. News and World Report* March 11, 2002.

(a)

x	-2	-1	0	1	2
$f(x)$	1	3	9	27	81

(b)

x	-2	-1	0	1	2
$f(x)$	$\frac{1}{2}$	$\frac{\sqrt{2}}{2}$	1	$\sqrt{2}$	2

(c)

x	-2	-1	0	1	2
$f(x)$	0.08	0.4	0	0.4	0.08

(d)

x	-2	-1	0	1	2
$f(x)$	0.309	0.555	1	1.8	3.24

(e)

x	0	2	4	6	8
$f(x)$	3	12	48	192	768

14. One of the following involves an exponential function and one involves a linear function. Identify which is which.

    (a) Find the cost of gasoline if the price is $1.23 per gallon.

    (b) Find the population of a city that is increasing at a rate of 3.5% per year.

15. Explain what is wrong with the following statement:

    > Since college tuition is increasing at a constant rate of 7.9% per year, and I know that linear functions have a constant rate of change, I can conclude that college tuition is increasing linearly.

16. On December 10, 1999 the *Dear Ann Landers* column ran a letter that claimed:[8]

    > When Elvis Presley died in 1977, there were 48 professional Elvis imper-sonators. In 1996, there were 7,328. If this rate of growth continues, by the year 2012, one person in every four will be an Elvis impersonator.

    (a) According to this claim, why would the Elvis impersonator function be expo-nential?

    (b) Find the "Elvis impersonator" exponential function that fits the 1977 and the 1996 data. Let your input be *years since 1977*.

    (c) Using your function, find the number of predicted Elvis impersonators in 2012.

    (d) The population in the United States is estimated to be 339,630,000 in 2012. Given this, does the claim that one person in every four will be an Elvis impersonator seem reasonable? Explain.

---

[8] *The Holland Sentinel* 10 Dec. 1999, A12.

17. For each of the following, find the equation of the exponential function that contains that pair of points.

    (a) (0, 1) and (1, 10)

    (b) (0, 2) and (2, 50)

18. Let $y = f(x)$ be an exponential function whose growth factor is 3. Suppose $f(3) = 108$.

    (a) For what value of x would $f(x) = 36$?

    (b) For what value of x would $f(x) = 324$?

19. Each of the following is an exponential function with a vertical shift, $f(x) = ba^x + c$. Give the equation.

    (a) $f(0) = 3$, $f(1) = 19$, and the horizontal asymptote is $y = 1$.

    (b) $f(0) = 0$, $f(1) = 1$, and the horizontal asymptote is $y = -1$.

    (c) $f(0) = 5$, $f(1) = 2.75$, and the horizontal asymptote is $y = 2$.

20. For each of the following, give the equation of the new exponential function.

    (a) $f(x) = 4^x$. Multiply the input by 3.

    (b) $f(x) = 2(\frac{1}{3})^x$. Add 1 to the input.

    (c) $f(x) = 10(3)^x$. Multiply the output by 2.

21. In Example 5, we discovered the African Elephant population has a growth factor of 0.95 per year. Suppose an error was made in population counts between 1980 and 1990, and the growth factor was actually 0.93.

    (a) What would be the new exponential function if input was years since 1980?

    (b) What would be the physical meaning of the new growth factor?

    (c) What would be the projected population in 2000? How does this compare to the number found in Example 5?

22. In Example 6, find the equation for the exponential function, $y = S(t)$, that represents the employees annual salary if he would have chosen the 2.3% cost of living increase. Let t be years since 1997.

23. In 2003 the average cost of electricity per 500 KWH (Kilowatt Hour) is $48.70. Using the function $V(p, t) = p(1.032)^t$, what do you predict the cost of electricity per 500 KWH to be in 2050?

24. In 1980, the average cost for a 16 ounce bag of potato chips was $2.10. Using the function, $V(p, t) = p \cdot 1.032^t$, answer the following questions.

    (a) Estimate the price for a 16 ounce bag of chips in 2001.

    (b) Estimate the price for a 16 ounce bag of chips in 1963.

Exponential Functions: Activities and Class Exercises

1. **Chicken Bacteria.** The information given in the table below was found on the label of a package of chicken.

Bacteria Count on Chicken
(per square centimeter when refrigerated at 40° F)

Time	Bacteria	Condition
Day 0	360	OK
Day 1	5,800	OK
Day 2	92,000	Fair
Day 3	1,475,000	Poor
Day 4	23,600,000	Odor
Day 5	377,500,000	Slime

It was also noted on the label that at 40° F, the bacteria double in number every 6 hours. Since the bacteria are growing at a constant rate, this is an exponential function. Let $B(t) = ba^t$ be the number of bacteria on the chicken after t days.

(a) Do you expect the value of a to be greater than 1 or less than 1? Why?

(b) Find reasonable values for b and a using the data in the table. [Remember that b is the y-intercept and a is the growth rate.]

(c) If the bacteria double every 6 hours, what will be the growth factor for one day? Why? How does this compare to your value for a?

2. **Heads or Tails.** You need a cup containing 100 pennies. Cover the cup with your hand, shake the cup to mix up the pennies, and then pour the pennies out on a table. Remove all the pennies landing heads up and count the remaining pennies. Record this number as the number of pennies remaining after the first trial. Put the remaining pennies back in the cup and repeat this process and record the number of pennies remaining after the second trial. Continue to repeat this process until all the pennies have been removed.

(a) Complete a table where the input is the number of trials and the output is the number of pennies remaining. Include trial 0 with 100 pennies remaining.

(b) Produce a scatterplot of your data.

(c) If the probability of a coin landing heads up is 0.5 and we start with 100 coins, what exponential function should theoretically model this situation where the input is the number of trials and the output is the number of coins remaining after all the heads are removed?

(d) Using exponential regression on your calculator or software, find the exponential function that fits the data that you found by tossing the coins. In doing so, do not include your last point that has 0 for an output since this will cause an error when your calculator tries to fit an exponential equation.

(e) Which function, the theoretical one or the one you found using your calculator, fits the data better? Explain how you determined your answer.

3. **Indy 500.** The following table gives the fastest lap speed of the Indy 500 that particular year.[9] Enter these data into your calculator where time is given in *years since 1950*. Use these data to answer the following questions.

Years	Racing Speed MPH
1951	133.81
1952	135.14
1953	135.87
1954	140.54
1955	141.35
1956	141.42
1957	143.43
1958	144.30
1959	145.42
1960	146.13
1961	147.59
1962	148.30
1963	151.54
1964	157.65
1965	157.51
1966	159.18
1967	164.93

Years	Racing Speed MPH
1968	168.67
1969	166.51
1970	167.79
1971	174.86
1972	187.54
1973	186.92
1974	191.41
1975	187.11
1976	186.03
1977	192.68
1978	193.92
1979	193.22
1980	190.10
1981	196.94
1982	200.54
1983	197.51
1984	204.85

Years	Racing Speed MPH
1985	204.94
1986	209.15
1987	205.11
1988	209.52
1989	222.47
1990	222.57
1991	222.18
1992	229.12
1993	214.81
1996	236.10
1997	215.63
1998	214.75
1999	218.88
2000	218.49
2001	219.830
2002	226.50

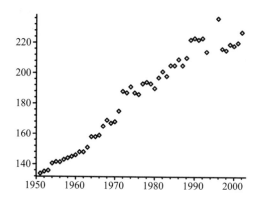

(a) Does this data have a positive or a negative association? Explain.

(b) Using your calculator, find the equation of the line that best fits this data. Let $y = L(c)$ denote the equation of your line.

    i. Use this equation to estimate the fastest lap speed of the car that won the Indy 500 in 1998.

    ii. The actual fastest lap speed in 1998 was 214.70 (by Tony Stewart on Lap 19). Explain why it was impossible for your linear equation to predict this particular speed.

    iii. Use this equation to predict the fastest lap speed of the car that will win the Indy 500 in 2030. Does this answer seem realistic? Explain.

    iv. What is the correlation for your line? What does this mean?

(c) Using your calculator, find the equation of the exponential function that best fits this data. Let $y = E(c)$ denote the equation of your exponential function.

    i. Use this equation to predict the fastest lap speed of the car that won the Indy 500 in 1998. How does this compare to the speed given by your linear equation?

    ii. Use this equation to predict the fastest lap speed of the car that will win the Indy 500 in 2030. Does this answer seem realistic? Explain.

    iii. What is the correlation for your exponential function? What does this mean?

(d) Do you think the speed of cars is increasing linearly or exponentially? Explain.

(e) What do you think happened in 1997?

4. **Population.** Population growth is a concern of many people around the world because of its impact on the environment, the food supply, and other limited resources. While this growth is very difficult to predict, particularly long-term growth, it is nonetheless necessary to make predictions in order to plan for the future. Mathematical models are often used to make reasonable predictions. These models can be quite complicated since population growth patterns are dependent upon many variable factors. However, relatively accurate information can be generated from simple mathematical models, particularly when their use is restricted to a short time period.

To help understand how growth rates affect the size of a population, we will compare population growth rates from three different countries. A document produced by the United Nations Population Fund gives an annual growth rate of approximately 1% for Ireland, a growth rate of approximately 2% for Mexico, and a growth rate of approximately 3% for Ethiopia from 1990 to 1995. (For this period, the United States had a growth rate of 0.71%.)[10] These countries range in population from about 4 million for Ireland to about 100 million for Mexico. For comparison purposes, assume that each of these countries had an initial population of 50 million.

---

10 *United Nations Population Fund.* "Population and the Environment: The Challenges Ahead," Banson Productions, 1991. 39–43.

(a) Each of the rates given are growth *rates*, not growth factors. A growth rate is used to determine the additional population, not the total population. This is similar to interest rates being used to determine additional money added to your account, not the total money in your account.

   i. Convert each of the growth rates to a growth factor.

   ii. Give an exponential function that models the population growth for each of the three countries, assuming the growth factor is constant.[11]

(b) Use the exponential functions you found in part (a)(ii) to complete the following table.

Country	Growth Rate	Population after 10 years	Population after 100 years
Ireland	1%		
Mexico	2%		
Ethiopia	3%		

(c) Graph the population for these three countries, on the same set of axes, with time on the horizontal axis and population on the vertical axis. First use a domain of 0 to 10 years, then on another graph, use a domain of 0 to 100 years.

(d) Describe the differences and similarities between the graphs with the 10-year domain and those with the 100-year domain. How are the shapes of the graphs in these two windows related to the resulting differences in populations of the three countries after 10 years and after 100 years?

(e) While the population of Ethiopia in 1990 was about 50,000,000, the population of Mexico was about 90,000,000. If the growth rates of these two countries stay constant, use the graphical features of a graphing calculator to estimate the year that Ethiopia's population would exceed that of Mexico.

5. **Carbon Dioxide.** Many people are concerned about the amount of carbon dioxide in the atmosphere. The table on the next page approximates the amount of CO_2 in the atmosphere for various years. The units are in giga-tons (one giga-ton is equal to 1,000,000,000 metric tons).[12]

(a) Using your calculator, find the equation of the line that best fits this data. Let $y = L(c)$ denote the equation of your line.

(b) What is the correlation for your line? What does this tell you about your line?

(c) Using your calculator, find the equation of the exponential function that best fits this data. Let $y = E(c)$ denote the equation of your exponential function.

---

[11] In reality, population growth rates do not stay constant. We will assume they do in order to simplify this problem.

[12] *Trends: Atmospheric Carbon Dioxide,* 16 March 2004,
<http://cdiac.esd.ornl.gov/ftp/maunaloa-co2/maunaloa.co2>.

Years	Amount of CO_2 in the atmosphere	Years	Amount of CO_2 in the atmosphere
1958	666.95	1980	718.28
1959	669.39	1981	721.33
1960	671.85	1982	722.92
1961	672.40	1983	726.93
1962	673.33	1984	732.35
1963	675.69	1985	732.67
1964	676.11	1986	736.21
1965	677.55	1987	740.28
1966	681.22	1988	746.37
1967	681.88	1989	748.66
1968	684.63	1990	753.00
1969	687.73	1991	756.06
1970	689.13	1992	757.07
1971	691.71	1993	760.50
1972	693.54	1994	764.41
1973	699.88	1995	768.67
1974	701.02	1996	772.42
1975	702.25	1997	774.81
1976	704.20	1998	780.92
1977	710.65	1999	784.46
1978	711.73	2000	786.82
1979	715.29	2001	790.00

(d) What is the correlation for your exponential function? What does this tell you about your exponential function?

(e) Which function, linear or exponential, is a better fit for the data? Why?

(f) According to the book *Global Environment*[13] whenever the atmospheric CO_2 doubles, the global mean earth surface temperature increases by 1.5° to 4.5°C.

---

[13] Berner, Elizabeth, Berner Kay, *Global Environment: Water, Air, and Geochemical Cycles*, Prentice Hall, 40.

This would mean that the winters in North America would be 4° to 6°C warmer.

    i. A change of 4° to 6°C is comparable to how many degrees of change in degrees Fahrenheit? [Note: $F = 1.8C + 32$.]

    ii. According to your linear function, L, when would the levels of CO_2 in the atmosphere be at double the 1958 levels?

    iii. According to your exponential function, E, when would the levels of CO_2 in the atmosphere be at double the 1958 levels?

  (g) Should we be concerned about the level of CO_2 in the atmosphere? Why or why not?

6. **Tuition versus Inflation.** College students dread getting the tuition bill for the following school year because the bill is expensive and always higher than the previous year. Many colleges contend that they are only adjusting their tuition in line with inflation. The following table shows the tuition charges for a midwestern liberal arts college and annual inflation rates from 1985 to 1999.[14]

Year	Tuition	% Consumers Price Index	Adjusted Tuition
1984–1985	$5,756	1.00	
1985–1986	$6,244	1.04	
1986–1987	$6,742	1.05	
1987–1988	$7,242	1.10	
1988–1989	$7,890	1.14	
1989–1990	$8,520	1.20	
1990–1991	$9,366	1.27	
1991–1992	$10,022	1.31	
1992–1993	$10,722	1.35	
1993–1994	$11,472	1.38	
1994–1995	$12,275	1.42	
1995–1996	$13,234	1.46	
1996–1997	$14,220	1.51	
1997–1998	$14,788	1.53	
1998–1999	$15,380	1.56	

---

[14] *Global Financial Data,* 16 March 2004, <http://www.globalfindata.com>.

(a) Complete the table by dividing the tuition for each year by its corresponding consumers price index. This will give the tuition adjusted for inflation.

(b) Using the academic year 1984–85 as year 0, 1985–86 as year 1, and so on, make a scatterplot of the year versus the tuition and a scatterplot of the year versus the adjusted tuition. Graph these two scatterplots on the same set of axes. Use different symbols (dots or crosses for example) or different colors for your two scatterplots.

(c) Based on your scatterplots, is tuition increasing faster than inflation? Explain.

(d) Using your calculator, find the equation of an exponential function that models each of these data sets: tuition versus year and adjusted tuition versus year.

(e) What is the growth factor for each equation? Describe what these numbers mean in terms of tuition and year.

(f) Predict the cost of tuition in 2010.

7. **Prison Populations.** The 1998 United States prison report stated that the number of imprisoned adults had more than doubled in the last 12 years, making our country second only to Russia as the country with the highest rate of incarceration at that point in time. (In the last half of 2000, this trend finally came to an end with the United States experiencing the first measured decline in prison populations since 1972.)[15]

Years	Number of Persons in Jail or Prison	Years	Number of Persons in Jail or Prison
1980	503,586	1991	1,219,014
1981	556,814	1992	1,295,150
1982	612,496	1993	1,369,185
1983	647,449	1994	1,476,621
1984	682,764	1995	1,585,586
1985	744,208	1996	1,646,020
1986	800,880	1997	1,743,643
1987	858,687	1998	1,816,931
1988	951,329	1999	1,893,115
1989	1,138,935	2000	1,933,503
1990	1,148,702		

---

[15] *Bureau of Justice Statistics*, 16 March 2004, <http://www.ojp.usdoj.gov/ bjs/glance/tables/corr2tab.htm>.

(a) Use the previous table to answer the following questions.

    i. Use your calculator to find the exponential equation that best fits this data. Let your input represent years since 1980.

    ii. What is your correlation? What does this tell you?

    iii. What would be your exponential function modeling prison population growth if you found your function by only using the two points (1980, 503,586) and (2000, 1,933,503)? [Note: Convert these to (0, 503,586) and (20, 1,933,503) before finding your function.] Comparing this to your answer from part (i), what can you conclude?

(b) Use the table for the United States population to answer the following questions.[16]

Years	Population of United States
1980	227,224,681
1981	229,465,714
1982	231,664,458
1983	233,791,994
1984	235,824,902
1985	237,923,765
1986	240,132,887
1987	242,288,918
1988	244,498,982
1989	246,819,230
1990	249,438,712

Years	Population of United States
1991	252,127,402
1992	254,994,517
1993	257,746,103
1994	260,289,237
1995	262,764,948
1996	265,189,794
1997	267,743,595
1998	270,298,524
1999	272,330,000
2000	274,634,000

    i. Use your calculator to find the exponential equation that best fits this data. Let your input represent years since 1980.

    ii. Compare the growth factor from the prison populations and the one for the United States's population. What does this imply? Explain.

    iii. Use your exponential function to estimate the U.S. population in 2010. Use your exponential function from (a)(i) to estimate the prison population in 2010.

    iv. According to this, what percentage of Americans will be imprisoned in 2010? Does this seem reasonable? Explain.

---

[16] *US Historical Population by Year*, 16 March 2004, <http://www.npg.org/facts/us_historical_pops.htm>.

8. **Cooling.** If you left a hot cup of coffee out on the counter for a while, you will end up with a lukewarm cup of coffee. Objects that are heated and then left to cool can be modeled with an exponential function. In this activity, we will heat a temperature probe using a cup of hot water and then measure its temperature for 90 seconds after it is removed from the cup.

   (a) Cooling is modeled mathematically with an exponential function. Before performing the cooling experiment, predict whether the growth factor (base) of the exponential function will be less than 1 or greater than 1. Justify your answer.

   (b) Perform the cooling experiment, being sure to record the beginning and ending temperatures for your data, the temperature you are using as room temperature, the exponential function found by the calculator, and the correlation, r.

   (c) Looking at the correlation, what do you know about the exponential function?

   (d) Sketch your time-temperature graph.

   (e) What is the growth factor (base) for your exponential function? SPECIFICALLY, what is this telling you about the temperature of the probe? [Note: It is not sufficient to say "the growth factor tells me the temperature is decreasing."]

Directions for Using the CBL with Exponential ACT

- You must first load the **CoolTemp**[17] program onto your TI-83 calculator. Only one member of your group needs to have this program.

- Plug the temperature probe into Channel two (top of the CBL) and connect the CBL to your calculator (using the small black cord).

- Choose the **CoolTemp** program.

- BEFORE starting to collect data, put the temperature probe into the cup of hot water. It should be in the water for approximately one minute before you start collecting data.

- Press ENTER until it says to start collecting data. As soon as you begin collecting data, remove the temperature probe from the water and hold it in the air. The CBL will collect data every second for 90 seconds.

- When the CBL is finished collecting data, use TRACE to find the starting temperature and ending temperature.

- If you are pleased with your graph, press Clear until you get a blank screen.

- The time data is stored in **L2** and the temperature data is stored in **L4**. To get an equation for the cooling, we have to subtract the room temperature from the temperature data. This is because we need to shift the function so that it "levels off" at the x-axis. To shift the function, we subtract the room temperature. Assume the

---

[17] This program can be downloaded at <http://www.math.hope.edu/swanson/calculator/programs.html>

room temperature is one-half degree less than your final temperature reading. Subtract this from the data in List 4 by pressing:

$$\boxed{\text{L4}} - \text{room temp } \boxed{\text{STO}} \boxed{\text{L4}}$$

- To fit an exponential function to your time-temperature data, choose $\boxed{\text{STAT}}$, **CALC**, **0: ExpReg**. Your time data is in List 2, your modified temperature data is in List 4, and you want to store the function in **Y1** to make it easy to graph. So you need to press:

$$\boxed{\text{L2}}, \boxed{\text{L4}}, \boxed{\text{VARS}}, \text{ \textbf{Y-VARS, 1:Function, 1:Y1}}$$

- Record the correlation coefficient (the value of r).

- Graph the time-temperature data and the exponential function.

10

Logarithmic Functions

In the previous sections, we looked at linear and exponential functions. These functions are similar in that they have a constant change, either a rate that is added or a factor that is multiplied. Logarithmic functions are different. These functions do not have a constant rate. In fact, they do not even use the basic algebraic operations of addition, subtraction, multiplication, division, or raising to a power. Instead, logarithms describe magnitudes. In this section, we will give the definition of the logarithmic function, explore some of its properties, and consider the types of applications where logarithms are used.

Magnitudes

Let's start by reviewing the metric system, something that turns out to be a natural use of the logarithm function. Recall that there are 10 meters in a dekameter, 100 meters in a hectometer, and 1000 meters in a kilometer. Each of these units is ten times larger than its preceding unit. The term magnitude is used to describe this type of relationship. If we use meters as our base, we would say that a kilometer has magnitude 3 since 10^3 meters = 1000 meters equals one kilometer. Similarly, a hectometer has magnitude 2 and a dekameter has magnitude 1. We could also say a meter has magnitude 0, since 10^0 meter = 1 meter.

Magnitude is defined in the dictionary as the "greatness of size" of something. For our purposes, we will define the **magnitude** of a number a as the exponent b such that $10^b = a$. Magnitudes can be defined using other numbers besides powers of ten, but we will choose 10 since our number system is based on powers of 10. Notice the relationship between magnitudes and exponents. A magnitude is really just another word asking for the exponent with a given base.[1] With this definition of magnitude, the magnitude of the number 10,000 is 4 since $10,000 = 10^4$. This is similar to the way we defined magnitude

---

[1] Recall that for c^d, the value of c is called the *base* and the value of d is called the *exponent*.

when looking at the metric system. This is because both our definition of magnitude and the metric system are based on powers of 10. It is pretty clear that the magnitude of 10 is 1 and the magnitude of 100 is 2. What, then, is the magnitude of 50? Since 50 is about half way between 10 and 100, it is reasonable to start with a guess of 1.5. We can check this by going back to our definition of magnitudes and using our calculator to find that $10^{1.5} \approx 31.6$. This is too low so we will increase our guess to 1.6. Since $10^{1.6} \approx 39.8$, this guess is still too low. Trying 1.7, we find $10^{1.7} \approx 50.1$. Therefore, the magnitude of 50 is approximately 1.7. Notice that magnitude is not a linear function. The rate of change between $(10, 1)$ and $(100, 2)$ is $\frac{2-1}{100-10} \approx 0.01111$. However, the rate of change between $(10, 1)$ and $(50, 1.7)$ is $\frac{1.7-1}{50-10} = 0.0175$. Since the rate of change is not constant, finding magnitudes is not a linear function.

Example 1.   Estimate the magnitude of 45,876 to two decimal places.

Solution.   Since 45,876 is between 10,000 (magnitude 4) and 100,000 (magnitude 5), it will have a magnitude between 4 and 5. To estimate the magnitude to two decimal places, we use a "guess and check" method.

Guess	4.60	4.65	4.66	4.67
Check	$10^{4.60} \approx 39,811$	$10^{4.65} \approx 44,668$	$10^{4.66} \approx 45,709$	$10^{4.67} \approx 46,774$

Since 45,876 is closer to $10^{4.66}$ than to $10^{4.67}$, its magnitude is approximately 4.66. ∎

There is a similarity between the magnitude of a number and writing that number in scientific notation. From Example 1, we saw that the magnitude of 45,876 was approximately 4.66. If we write 45,876 in scientific notation we find that $45,876 \approx 4.59 \times 10^4$. We see that the exponent on 10 is simply the whole number part of the magnitude of 45,876. Since both magnitudes and scientific notation are based on powers of 10, this is not surprising.

The Logarithm Function

The function whose input is a number and whose output is its magnitude is called the **logarithm** function, or more simply, the **log** function. The logarithm function we will use will always have base 10. Recall that *base* refers to the number you are raising to a power. So $y = \log x$ means $10^y = x$. In other words, y is the exponent which, when 10 is raised to that power, gives us x. The output of the logarithm function can be called either the magnitude or the logarithm. Table 10.1 shows the logarithm (or magnitude) for some numbers between 0 and 10.

Note that the points $(1, 0)$ and $(10, 1)$ are included in Table 10.1. This implies that $\log(1) = 0$ and $\log(10) = 1$. This is because the logarithm function is telling us the magnitude (or exponent) of the number. So $\log(1) = 0$ is the same as $1 = 10^0$ while $\log(10) = 1$ is the same as $10 = 10^1$.

The graph in Figure 10.1 is the logarithm (or magnitude) function, $y = \log x$, for $0 < x \leq 10$. Notice that the graph is increasing and concave down. It is increasing because the larger the number, the larger its magnitude. It is concave down because the

TABLE **10.1**

The logarithm function for values of x where $0 < x \leq 10$.

x	0.1	0.5	1	2	3	4
$\log x$	-1	≈ -0.30	0	≈ 0.30	≈ 0.48	≈ 0.60
x	5	6	7	8	9	10
$\log x$	≈ 0.70	≈ 0.78	≈ 0.85	≈ 0.90	≈ 0.95	1

rate of change is decreasing. We can see this by looking at Table 10.1. Notice that when the input increased from 1 to 2, the output increased by 0.30. When the input increased from 9 to 10, the output increased by only 0.05.

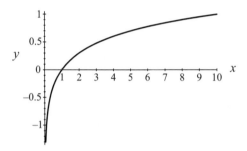

FIGURE **10.1**

The logarithm function using the domain $0 < x \leq 10$.

The graph of $y = \log x$ will continue to be concave down for $10 \leq x \leq 100$. (See Figure 10.2.) Since $\log 10 = 1$ and $\log 100 = 2$ (i.e. $10^2 = 100$), the increase in the output is one unit—the same as the increase in the output between $\log 1 = 0$ and $\log 10 = 1$. Yet for $1 \leq x \leq 10$, the total change in the input is $10 - 1 = 9$ units and the change in the output is one unit. However, for $10 \leq x \leq 100$, the total change in the input is $100 - 10 = 90$ units while the change in the output is again only one unit. The same

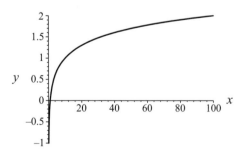

FIGURE **10.2**

The logarithm function on a domain of $0 < x \leq 100$.

increase in height (or change in outputs) is spread over a much larger interval. This causes the graph to be concave down.

In general, the logarithm function's graph will be steep at the beginning, but will quickly become flatter and flatter as your input increases. The logarithm function continues to grow, although *very* slowly.

Properties of the Logarithm Function

Basic Properties

An inspection of Figures 10.1 and 10.2 reveals several properties of the logarithm (or magnitude) function. Some of these properties of $y = \log x$ are:

- $\log(1) = 0$.
- $\log x$ is negative for $0 < x < 1$ and positive for $x > 1$.
- $y = \log x$ has a domain of $x > 0$ and a range of all real numbers.
- The graph of $y = \log x$ is increasing and concave down.

These properties are a result of the properties of exponents. This is true because logarithms are merely magnitudes or exponents. For example, the magnitude of 1 is 0 because $10^0 = 1$. This is analogous to saying a meter has magnitude 0. The magnitude of something less than one (but greater than 0) is negative. This is because you have to take the reciprocal of 10 to get a number smaller than 1. For example, $\log(\frac{1}{100})$ means "find the exponent, y, such that $10^y = (\frac{1}{100})$." Since $(\frac{1}{100}) = \frac{1}{10^2} = 10^{-2}$, we have $\log(\frac{1}{100}) = -2$. This is similar to saying a centimeter has magnitude -2. The magnitude of anything greater than 1 is positive. If you want to raise 10 to a power to get a number greater than one, your exponent will be positive. For example, the magnitude of $1000 = \log 1000 = 3$.

So far we have looked at finding magnitudes of positive numbers. What about finding magnitudes of negative numbers? Notice that in Figures 10.1 and 10.2, $y = \log x$ does not exist for negative values of x. This is because the logarithm of a negative number is undefined. For example, what is the exponent 10 needs to be raised to in order to get -100? There is not one. Ten raised to *any* power will always be positive. Similarly, $\log 0$ is undefined. There is no exponent, p, such that $10^p = 0$. In terms of the metric system, there is no unit to describe a length of zero or a negative length. This is why the domain of the log function is restricted to the set of positive real numbers.

Adding Logarithms

It was mentioned earlier that the output of the logarithm function increases by one unit as the input increases from 1 to 10. The output again increases by one when the input increases from 10 to 100. This idea was helpful when observing the shape of the graph of the logarithm function. Because the logarithmic function is defined as magnitudes, the output of the logarithm function will always increase by 1 whenever the input changes by a factor of 10. For example,

$$\log(2) \approx 0.3010 \text{ since } 10^{0.3010} \approx 2$$

and

$$\log(20) \approx 1.3010 \text{ since } 10^{1.3010} = 10 \cdot 10^{0.3010} \approx 10 \cdot 2 = 20.$$

Multiplying a number by 10 increases the magnitude by one unit. Symbolically, if $\log c = q$, then $\log 10c = q + 1$.

Let's do another example. Let a and b be numbers such that $\log(a) = 3$ and $\log(b) = 4$. This means $a = 10^3$ and $b = 10^4$. Notice that $\log(a) + \log(b) = 3 + 4 = 7$. What input would give an output of 7? It would be 10^7, the *product* of the original two inputs, 10^3 and 10^4.

In general, we can say that adding the outputs of the logarithm function corresponds to multiplying the inputs. Symbolically, we summarize this property as:

- $\log(a) + \log(b) = \log(ab)$.

Even though this may seem like a new concept, it is merely a consequence of the properties of exponents. Whenever you multiply two numbers with exponents that have the same base, you add the exponents. Symbolically, if $\log a = p$ and $\log b = q$, then $10^p = a$ and $10^q = b$. So $ab = 10^p 10^q = 10^{p+q}$. This means $\log(ab) = p + q = \log(a) + \log(b)$.

Logarithms were originally developed for computational purposes (see the Historical Note later in this section). This property of converting a multiplication problem to an addition problem is one of the reasons logarithms were so useful before calculators became readily available. Addition of large numbers is much easier than multiplication! If we wanted to use logarithms to multiply two very large numbers, we would use a log table (these were readily available in the pre-calculator days) to find the logs of our two numbers and then add the two logarithms. We would take this sum and, reading our log table "backwards," find the result of our multiplication without actually doing any multiplication.

Multiplying Logarithms by Constants

The property of being able to add the logarithms of numbers to find the logarithm of a product leads to another property of logarithms. Let's start by using $\log(a) + \log(b) = \log(ab)$ to find $\log a^2$. Since

$$\log(a^2) = \log(aa),$$

we know

$$\log(aa) = \log(a) + \log(a)$$
$$= 2\log(a).$$

Similarly,

$$\log(a^3) = \log(a) + \log(a^2)$$
$$= \log(a) + 2\log(a)$$
$$= 3\log(a).$$

In general, for all positive integer values of n, this pattern gives us

$$\log(a^n) = n\log(a).$$

This property shows that, for positive integer exponents, the logarithm of a number raised to an exponent is the same as the value of the exponent multiplied by the logarithm of the number. However, this property is true for all real exponents as well. Again, this is a consequence of the properties of exponents. In particular, it's a consequence of the property "when you raise an exponent to an exponent, you multiply the exponents." To illustrate this symbolically, suppose $\log(b) = p$. Then $b = 10^p$. Therefore, $b^r = (10^p)^r = 10^{pr}$. So $\log(b^r) = pr = r \log(b)$. In summary,

- $\log(a^r) = r \log(a)$ for any real number, r, and any positive number, a.

This second property of logarithms again made logarithms a valuable tool for doing complicated calculations before the era of calculators.[2] Finding the value of a number raised to an exponent r was much easier using logarithms. Instead of repeated multiplication, you could merely multiply the log of the number by r. As an example, finding the value of 12^{10} by hand would be difficult. But, using logarithm tables, we look up $\log 12$ and find that it is approximately 1.079. We know that $\log 12^{10} = 10 \log 12$ so multiplying 1.079 by 10 gives us 10.79. The antilog of 10.79 (that is, the number whose log is 10.79) is found in a logarithm table as $\approx 6.166 \times 10^{10}$. This implies that $12^{10} \approx 6.166 \times 10^{10}$. Logarithms also gave a way of finding things such as $2^{1/2} = \sqrt{2}$ by using $r = \frac{1}{2}$.

Historical Note[3]

John Napier, the man who is usually credited with the invention of logarithms, was not a professional mathematician; he was a Scottish laird who wrote on many different topics while managing his large estates. Napier was seeking a procedure to make computations involving large numbers easier. He published his system of logarithms in 1614. This book, whose title translates as *A Description of the Marvelous Rule of Logarithms*, was immediately admired by Henry Briggs, a professor of geometry at Oxford. In 1615, Napier and Briggs had a meeting and agreed that 10 should be used as the base for his logarithms. In 1617, the year of Napier's death, Briggs published a table of logarithms of the numbers 1 through 1,000 each to fourteen places! This table was extended in 1624 to include the logarithms (again to fourteen places) of the numbers 1 through 20,000 and 90,000 through 100,000.

The English word logarithm comes from two Greek words: *logos* meaning "reckoning or computing" and *arithmos* meaning "number."[4] Logarithms were, from their inception, meant to be a way to reckon or calculate numbers. It was later that many other uses for Napier's interesting discovery, the logarithm function, were found.

---

[2] Slide rules, a simple computational device popular before calculators, used logarithms for doing many calculations.

[3] Boyer, Carl B, *History of Mathematics,* Princeton University Press, Princeton, pp. 342–345.

[4] Schwartzman, Steven, *The Words of Mathematics: An Etymological Dictionary of Mathematical Terms Used in English,* Mathematical Association of America, Washington, DC, p. 128.

Why Use Magnitudes?

Many scientific measurements are in terms of magnitudes. For example, the pH scale in chemistry, the way earthquakes are measured, distances to the stars, and sound intensity are all measured in terms of magnitudes and therefore use a logarithmic scale. One reason for this is that it is easy to distinguish between magnitudes of change while it is often difficult to distinguish change measured in differences. An illustration of this is shown in Figure 10.3.

Figure 10.3 contains 12 boxes of dots. The number of dots in each box is indicated at the bottom of the box. Notice that each box in the second row contains one more dot than each corresponding box in the first row and that each box in the third row contains 10 more dots than each corresponding box in the first row. As you look down the first column of boxes, it is very easy to see the increase in the number of dots. In the second column this same increase is more difficult to see. In the third and fourth columns this same increase is almost impossible to detect. However, by looking across any of the three rows, it is easy to see the increase in the number of dots per box. The increase across the rows goes from a magnitude of 0 ($10^0 = 1$), to a magnitude of 1 ($10^1 = 10$), to a magnitude of 2 ($10^2 = 100$), and finally to a magnitude of 3 ($10^3 = 1000$). Your eye perceives the change of magnitudes better than it perceives the change in terms of individual dots, especially when the number of dots is large.

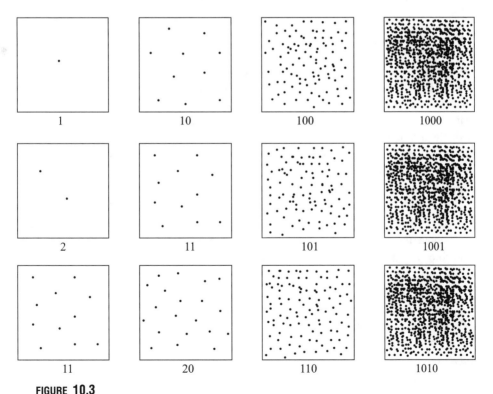

FIGURE 10.3
These boxes help visualize the difference between changes in magnitudes and differences.

Another situation modeled with magnitudes is the way you perceive sound. It has been suggested that we do not perceive changes in sound intensity linearly, but rather logarithmically. The smallest change in sound intensity that the human ear can detect is about 26%. This corresponds to a change of one decibel. The volume control on your stereo or television compensates for the human ear's nonlinearity by using a "logarithmic-taper potentiometer." This device is designed to give the impression of a linear increase as the volume is raised.[5] Look again at the dots in Figure 10.3. Let's imagine that these represent sound. If your stereo had a top volume of 1000 dots and your volume control was linear, then turning the volume dial one quarter of the way would be the equivalent of going from 1 dot to 250 dots in that quarter turn. You would go from no sound to a great deal of sound very quickly. This would represent a change of about 24 decibels.

Let's see why increasing our sound level from 1 to 250 is the same as an increase of about 24 decibels. Remember that a decibel was defined as a change in sound intensity of about 26%. Since a decibel is a 26% increase, the function modeling decibels has a growth rate and thus is an exponential function. An increase of 26% means the level of sound after an increase of one decibel is $1 + 0.26 = 1.26$ of the original sound level. So decibels are an exponential function with a growth factor of 1.26. An increase of 24 decibels is $(1.26)^{24} \approx 250$, the increase in the sound level when we turned our linear volume control. It is no accident that logarithms, which measure magnitudes, are closely related to exponential functions, which have a constant growth factor. Both of these functions are based on using exponents. For exponential functions, the exponent is the input. For logarithmic functions, the exponent is the output.

Let's return to the sound system with the linear volume control. The next quarter turn, from 250 dots to 500 dots, would still represent a change in sound. However, this turn would only represent a change of about 3 decibels since $(1.26)^{24+3} = (1.26)^{27} \approx 500$. The next quarter turn, from 500 to 750, would represent a change of only about 1.6 decibels. In the last quarter turn of our linear volume control, you would barely be able to hear any difference in sound level. Without the logarithmic-taper potentiometer, it would be difficult to control the volume since almost all of the change would occur in the first quarter-turn of the volume control. Your volume level would go from too soft to too loud very quickly. Notice how this is similar to the graph of the logarithm function which increases rapidly at first, and then increases more slowly.

The shape of the logarithm graph makes logarithms a good modeling function for behavior that increases rapidly at first, but then slows down dramatically. For example, think about the way you learn how to use a new piece of software. You quickly go from knowing nothing about that particular program to learning enough to start using the program. Your knowledge of the program increases quickly in a short period of time. However, soon your rate of learning about the program slows down. You continue to learn as you encounter new situations, but you only learn a few new things now and then. So your learning went from an initial rapid growth to a slow, gradual growth. This means that a modified logarithmic function would best model the process of learning to use the software.

---

[5] Gibilisco, Stan and Neil Sclater, editors, *Encyclopedia of Electronics, 2nd Edition*, TAB Professional and Reference Books, 1990.

Application: Decibels[6]

In the early 1900s, Bell Laboratories did a comprehensive study of human hearing. They wanted to develop a standard measurement of how loud we hear things. The initial idea was that they would increment the measurements by making each sound twice as loud as the previous. Each increment would be called a "Bel" out of respect for Alexander Graham Bell. The plan was to divide the Bel into 10 equal parts (the decimal system) and call each part a "decibel." A base reference was needed so that the measurement system could be used accurately. Bringing in extremely sensitive barometric pressure sensors, they measured the changes in the air pressure of the room. Using a 1 kHz tone, they found that the quietest sound their subject could hear was at 0.0002 micro-bars pressure. This is the point that Bell Labs referenced to 0 dB-SPL (Sound Pressure Level) which is commonly referred to as the "threshold of hearing." The "threshold of pain" is at approximately 200 micro-bars pressure. One of the more interesting findings during the study was that human hearing is not linear, meaning that twice the barometric pressure does not create a sound that we perceive as being twice as loud.

The equation for converting sound pressure level to decibels is

$$\text{decibels} = 20 \cdot \log \left(\frac{\text{sound pressure level}}{0.0002} \right). \tag{1}$$

Example 2.   Find the decibels for the threshold of pain. This is the amount of noise produced by a jack hammer, cymbal crash, or a jet plane's engine.

Solution.   The threshold of pain is given as 200 micro-bars pressure. We will convert this to decibels using Equation (1). We know our input is a sound pressure level of 200, therefore:

$$\text{decibels} = 20 \log \frac{200}{0.0002}$$
$$= 20 \log(1,000,000)$$
$$= 20 \times 6$$
$$= 120.$$

Painful noises are at 120 decibels. ∎

Let's look at the other end of the spectrum at sounds that are barely detectable. A sound that is at the threshold of hearing would be at 0.0002 micro-bars of pressure. This corresponds to 0 decibels.

Example 3.   Find the decibels for a sound that has twice the pressure as the threshold of hearing, i.e., 0.0004 micro-bars pressure.

Solution.   Once again we will use Equation (1) to get

---

[6] *dB-SPL*, 17 March 2004, <http://www.ptme.com/et/audio/reference/sound/dB-SPL.htm>.

$$\text{decibels} = 20 \log \frac{0.0004}{0.0002}$$
$$= 20 \log 2$$
$$\approx 20 \times 0.3010$$
$$\approx 6.$$

So doubling the pressure creates a sound that is approximately six decibels louder. ∎

When interpreting the sound level of six decibels, keep in mind that a normal person to person conversation is 60 decibels and a quiet bedroom at night is 30 decibels. When you look at the decibel differences between the threshold of hearing and the threshold of pain, remember these are *magnitudes* of the actual sound pressure levels.

Example 4. When the sound level of a television was measured, we found that it was about 40 decibels at the quietest setting and about 80 decibels at the loudest setting. By what factor did the pressure that is creating the sound increase?

Solution. We are given measurements of 80 decibels and 40 decibels, so we will convert these measurements into micro-bars to determine the difference in pressure. Using the equation for the sound pressure level, we want to find the sound pressure level for 40 decibels. We start by dividing both sides of our equation by 20.

$$40 = 20 \log \left(\frac{\text{sound pressure level}}{0.0002} \right)$$
$$2 = \log \left(\frac{\text{sound pressure level}}{0.0002} \right)$$

Since $2 = \log 100$,
$$\left(\frac{\text{sound pressure level}}{0.0002} \right) = 100.$$

Multiplying both sides of this equation by 0.0002 gives us

$$\text{sound pressure level} = 0.02.$$

Now using the equation for the sound pressure level, we want to find the sound pressure level for 80 decibels. Again, we start by dividing both sides of our equation by 20.

$$80 = 20 \log \left(\frac{\text{sound pressure level}}{0.0002} \right)$$
$$4 = \log \left(\frac{\text{sound pressure level}}{0.0002} \right)$$

Since $4 = \log 10,000$,

$$\left(\frac{\text{sound pressure level}}{0.0002} \right) = 10,000.$$

Multiplying both sides of this equation by 0.0002 gives us

$$\text{sound pressure level} = 2.$$

Therefore, 80 decibels is 2 micro-bars, and 40 decibels is 0.02 micro-bars. To determine the increase in pressure, we see that $\frac{2}{0.02} = 100$, so 80 decibels has 100 times the sound pressure of 40 decibels. ∎

Reading Questions for Logarithmic Functions

1. For each of the following, decide whether it would be best to make comparisons in terms of numbers (thereby using a linear scale) or in terms of magnitudes (thereby using a logarithmic scale). Justify your decision.

   (a) The body mass of animals ranging from insects to mammals.

   (b) The distribution of salaries for faculty at a college or university.

   (c) The distances to various stars.

   (d) The history of the world starting with when the Earth was formed.

   (e) The history of mankind starting with the earliest written records.

   (f) The net worth of companies in the United States.

2. What is meant by the *magnitude* of a number?

3. What is the magnitude of ten thousand? One million?

4. Estimate the magnitude of 2,984,200 to 2 decimal places.

5. In this section, it was mentioned that the magnitude function cannot be a linear function. Explain why this is true.

6. Estimate p to two decimal places.

   (a) Find p such that $10^p \approx 121$.

   (b) Find p such that $10^p \approx 16$.

   (c) Find p such that $10^p \approx 4900$.

7. What is the range of the graph of the logarithm function, $y = \log x$, for $1 \le x \le 10,000$?

8. Answer each of the following:

   (a) If $r < 0$, what is $\log r$?

   (b) If $\log r < 0$, what is r?

   (c) If $0 < r < 1$, what is $\log r$?

   (d) If $0 < \log r < 1$, what is r?

9. Why is it impossible to find a real number, p, such that $10^p = -8$?

10. Given that $\log 9 \approx 0.954$, find each of the following.

    (a) $\log 3$

    (b) $\log 81$

11. Given that $\log 4 \approx 0.602$, find each of the following.

    (a) $\log 400$

    (b) $\log 2$

    (c) $\log 16$

    (d) $\log 40$

12. Given that $\log 3 \approx 0.477$ and $\log 4 \approx 0.602$, find $\log 12$ without using a calculator.

13. Find a such that $\log a \approx 1.2$.

14. Why is measuring in magnitudes useful when trying to perceive the difference in large quantities?

15. What is the growth factor for the exponential function whose input is the number of decibels and whose output is the sound level?

16. In the example involving decibel level, we stated that if the sound level went from 500 to 750, the decibels would increase by less than 2. (The decibel level at 500 is approximately 27.) Show that this is true.

17. What do logarithm functions and exponential functions have in common?

18. Describe a physical situation (not mentioned in this section) for which a logarithm function might be a good model.

19. Knowing that the threshold of hearing is at 0.0002 micro-bars pressure and that the threshold of pain is at 200 micro-bars pressure, explain why using logarithms is a good way to measure sound.

20. The sound level in movies is set so that normal on-screen conversation with no music or special effects in the background is at 70 decibels. However, special effects (explosions, gunfire, etc.) can cause the sound level to reach 110 decibels.[7] By what factor was the pressure increased that is creating the sound?

Logarithmic Functions: Activities and Class Exercises

1. **Benford's Law.** In 1938 Dr. Frank Benford, a physicist at the General Electric Company, noticed that, given a list of numbers, more numbers started with the digit 1 than any other digit. Random numbers are equally likely to start with a $1, 2, 3, \ldots, 9$ yet that did not appear to be what really happened in lists of data. Dr. Benford

---

[7] Bondi, Nicole, "Hearing: Don't worry: Special effects won't affect ears," *The Detroit News*, July 30, 1997 <http://www.detnews.com/1997/metlife/9707/30/07300067.htm>, 2 Sept. 1998.

analyzed 20,222 sets of numbers including areas of rivers, baseball statistics, numbers in magazine articles, and street addresses. He concluded that the probability of the numeral d appearing as the first digit was

$$\log\left(1 + \frac{1}{d}\right)$$

In other words, according to Benford's Law, the probability that the first digit is a 1 is $\log(1 + 1) \approx 0.30$, the probability that the first digit is a 2 is $\log(1 + \frac{1}{2}) \approx 0.18$, etc. So about three-tenths of the numbers in a list typically start with the digit 1 (even though a random list of numbers should have only one-ninth of the numbers starting with the digit 1.)[8]

(a) Complete the following table.

d	1	2	3	4	5	6	7	8	9
$\log(1 + 1/d)$									

This table, according to Benford's law, gives the proportion of first digits that are 1s, 2s, etc.

(b) We looked at 60 retired baseball jersey numbers from the American League.[9] The following table indicates how many numbers started with a $1, 2, 3, \ldots, 9$.

First Digit of Jersey Number	1	2	3	4	5	6	7	8	9
Frequency	17	11	12	5	5	2	2	3	3

Do these numbers seem to conform to Benford's Law? Justify your answer by comparing a relative frequency distribution of the first digit of the jersey number to the distribution from part (a).

(c) We also looked at the populations in 1990 of counties in Michigan.[10] The following table indicates how many numbers started with a $1, 2, 3, \ldots, 9$. There are 83 numbers in all.

First Digit of Population Number	1	2	3	4	5	6	7	8	9
Frequency	26	17	10	4	9	2	7	5	3

Do these numbers seem to conform to Benford's Law? Justify your answer by comparing a relative frequency distribution of the first digit of the population number to the distribution from part (a).

---

[8] Browne, Malcom W. "Looking Out for No. 1." *New York Times*, 4 August 1998, p F4.
[9] *Baseball Almanac—The "Official" Baseball History Site*, 22 August 2002, <http://www.baseball-almanac.com>.
[10] Time Series of Michigan Population Estimates by County, 22 August 2002, <http://eire.census.gov/popest/data/counties/tables/CO-EST2001-07/CO-EST2001-07-26.php>.

(d) Using one page from a newspaper, list all of the numbers that appear. Do they seem to conform to Benford's Law? Justify your answer.

(e) Benford's Law is currently used in software designed to detect fraud and tax evaders. Explain how you think it does this and why this is likely to work.

2. **pH.** By definition, pH is the negative log of the hydrogen ion (H+) concentration. The hydrogen ion is the ion that makes acids acidic. Likewise, a basic solution is basic because of the hydroxide ion (OH-). Hydrogen and hydroxide concentrations are related to each other. The relationship is an equilibrium system where, as the concentration of hydrogen drops, the concentration of hydroxide will rise to compensate. In pure undisturbed water, this equilibrium will exist at a pH of 7 with the same concentration of each ion.

The formula for finding pH is

$$\text{pH} = -\log[H^+]$$

where H^+ is the hydrogen ion content in moles per liter of a given solution. The solution is said to be acidic if the pH is less than 7 and basic if the pH is greater than 7.

(a) The concentration of H^+ can range from about 0.1 moles per liter to about 10^{-14} moles per liter. Explain why this means that using logarithms is a good way to measure pH.

(b) Orange juice has a H^+ concentration of about 0.000316 moles per liter. What is its pH?

(c) Sea water has a H^+ concentration of about 0.000000005 moles per liter. What is its pH?

(d) Find the pH of a solution whose H^+ concentration is 2×10^{-7} moles per liter. Knowing that a solution with pH of 7 has a H^+ concentration of 10^{-7} moles per liter, complete the following sentence.

Doubling the acidity of a solution from 10^{-7} to 2×10^{-7} decreases the pH level by approximately _____ .

(e) Normal rain water has a pH of about 5.6. In 2002, the pH level of precipitation in western Michigan was about 4.67.[11] How many times more acidic was this precipitation than ordinary rain water?

3. **Seed Viability.** A biology professor conducted an experiment in which he buried mesh bags with seeds in the ground. At certain intervals, he removed seeds from the bags and planted them to determine if the seed was still viable. The following are the data he found along with a scatterplot of the data and a curve fitted to the data.

---

[11] *National Atmospheric Deposition Program (NRSP-3)/National Trends Networdk. (2004) Annual Data Summary,* 17 March 2004, <http://nadp.sws.uiuc.edu/nadpdata/ads.asp?site=MI53>.

Days	Avg. Proportion Viable (v)
0	0.88
203	0.72
498	0.5564
679	0.4503
1043	0.2866

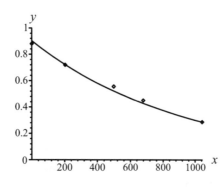

(a) Looking at the previous table of data along with the scatterplot, answer the following questions.

  i. What properties of the graph would lead you to conclude that this might be an exponential function?

  ii. Using your calculator or software, compute the exponential regression equation for the data points.

(b) Scientists often take the log of the output of data that follow an exponential pattern. This causes the data to look linear instead. The following is the graph of the linear regression line where the input is days and the output is the log(avg. proportion viable). Calculate the actual log(avg. proportion viable) values by completing the following table.

Days	v	$\log v$
0	0.88	
203	0.72	
498	0.5564	
679	0.4503	
1043	0.2866	

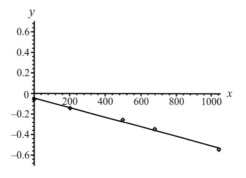

(c) The equation for the line is $y = -0.0004616x - 0.0448940$, where y is the log(proportion viable).

  i. Replace y in the equation by $\log y$ and then solve for y. (Hint: In rewriting this new log equation as an exponential, you get $y = 10^{-0.0004616x-0.0448940}$. Simplify this exponential equation.)

  ii. Compare your answer to part(a)(ii). What is the relationship between the slope of the line $y = -0.0004616x - 0.0448940$ and the growth factor of your exponential function?

  iii. What is the relationship between the y-intercept of your line $y = -0.0004616x - 0.0448940$ and the y-intercept of your exponential function?

4. **Earthquakes.** Consider a rubber band that is being stretched. The tension increases the amount of energy being stored in the rubber band. When the rubber band is pulled too far and snaps, the energy is released. The physical quantity known as *moment* computes the energy released times the distance the rubber band snapped back. This is similar to what occurs during an earthquake. You are probably accustomed to hearing magnitudes of earthquakes measured on the Richter Scale. In recent years, however, geologists have been using a more accurate scale called the Moment Magnitude.[12] The moment measures energy times distance. For an earthquake, this is calculated as the distance of the slip of the fault, d, times the area of the fault surface that slips, A, times the strength of the ground-up rock on the fault, μ, or $M_o = A \times d \times \mu$. The moment is converted into magnitude using the equation

$$M_w = \frac{2}{3} \log M_o - 10.7.$$

Because we are looking at magnitudes, a change of one unit does not mean "just a little bit more." To see this, suppose two earthquakes have moment magnitudes of 6 and 4. We can write the moment magnitudes equation as:

$$6 = \frac{2}{3} \log M_o - 10.7 \quad \text{and} \quad 4 = \frac{2}{3} \log M_o - 10.7.$$

Using algebra, we rewrite the equations as

$$25.05 = \log M_o \quad \text{and} \quad 22.05 = \log M_o.$$

Since we saw that $y = \log x$ means $x = 10^y$, these equations can be written as $M_o = 10^{25.05}$ for a magnitude of 6 and $M_o = 10^{22.05}$ for a magnitude of 4. To determine how many times larger the moment of the 6 magnitude quake is over the 4, we need to find the proportion of the two. In order to do this, we divide the larger magnitude by the smaller magnitude, which in this case is $\frac{10^{25.05}}{10^{22.05}} = 1000$. So a difference of 2 in the moment magnitude produces an earthquake 1000 times stronger.

(a) Why does the moment magnitude function involve a logarithm function?

(b) On March 28, 1964, one of the world's strongest earthquakes occurred in Prince William Sound, Alaska and had a magnitude of 9.2.[13] In San Francisco, another deadly earthquake occurred on April 18, 1906 and had a magnitude of 7.8. Determine how many times larger the moment was for the Alaskan quake than the San Francisco quake nearly 60 years before.

(c) On March 3, 2002, an earthquake with a magnitude of 7.4 killed nearly 1000 people in Afghanistan. A few weeks later another earthquake with a magnitude of 5.6 occurred. How many times larger was the moment of the first quake than the second quake?

---

[12] *USGS Earthquake Hazards Program-FAQ*, 17 March 2004, <http://earthquake.usgs.gov/faq/meas.html>.
[13] *USGS Earthquake Hazards Program: Largest Earthquakes in the United States*, 17 March 2004, <http://neic.usgs.gov/neis/eqlists/10maps_usa.html>.

(d) Subtract the magnitudes for the Alaskan quake and the San Francisco quake. Do the same for the quakes in Afghanistan. Compare the difference between the magnitudes and your answers to parts (b) and (c) regarding moments. What conclusions can you make about reading or viewing reports where the results are in magnitudes?

11

Periodic Functions

The average daily temperature for Kansas City changes from day to day, but does so in a cyclic pattern, changing from hot to cold and back to hot again and then repeating that cycle from year to year. The position of the second hand on a clock repeats itself every minute. The position of a child swinging on a swing or a person pacing back and forth is also repetitious. Repetitious behavior is modeled mathematically with periodic functions.[1] None of the functions we have discussed so far will describe these sorts of phenomena since linear, exponential, and logarithmic functions do not repeat the same set of outputs over and over again. In this section, we will show how periodic functions are defined, give situations that can be modeled with periodic functions, and give an introduction to two important periodic functions: the sine and cosine.[2]

What are Periodic Functions?

Imagine watching the second hand move around the face of a clock. The second hand returns to the same position every 60 seconds. Because this repetition occurs at regular intervals, this behavior is called periodic. For something to be periodic, it must not only repeat itself, but the repetitions must also be at regular intervals. A **periodic function** is one that gives the same output for every pair of inputs that are a fixed distance apart. Symbolically, we say f is periodic if for some real number p, $f(x) = f(x + p)$ for all x. The **period** of this function is the smallest value of p for which this relationship is always true, i.e., the period is the *minimum* fixed distance such that the outputs are always the same. For example, if the period of $y = f(x)$ was 4, then $f(x)$ and $f(x + 4)$ would be equal for *every* value of x. Figure 11.1 is an example of a periodic function with a period of 4. Notice that $f(0) = f(4) = f(8) = f(12) = 0$ while $f(1) = f(5) = f(9) = 1$.

---

[1] Magazines that are published at regular intervals are called *periodicals* for the same reason.
[2] Note: We will only be defining $\cos x$ and $\sin x$ as positions on the unit circle. We will not be using the right triangle definitions.

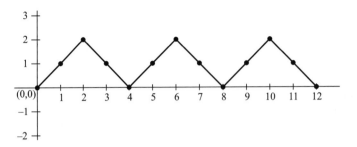

FIGURE **11.1**
The graph of a periodic function with a period of 4.

If a function is represented graphically, it is usually easy to determine whether or not it is periodic. Four graphs are shown in Figure 11.2. Graph (a) is not periodic. While it is going up and down on equal intervals, it does not give the same output on each of these intervals. Graph (b) is not periodic since the intervals on which it repeats are not all the same size. Graph (c) is a periodic function since it gives the same output for any two points a fixed distance apart. Even though graph (d) repeats itself on equal intervals, it is not the graph of a function.

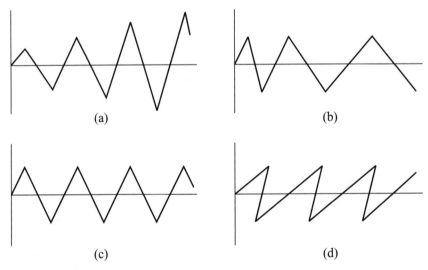

FIGURE **11.2**
Of these four graphs, only graph (c) represents a periodic function.

Graphing a Periodic Function

Consider the face of a clock centered on an xy-coordinate system. (See Figure 11.3.) The position of the tip of the second hand can be described in terms of its position on this coordinate system. We will describe this position in two ways. We will use one function

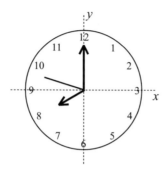

FIGURE 11.3
A clock centered on an xy-coordinate system.

to describe the *vertical* position of the point on the end of the second hand and another to describe the point's *horizontal* position. We will use time as an input for both of these functions. The horizontal axis of our graph will be labeled 0–120 seconds. [Note: We chose 120 seconds because it gives two cycles of the second hand.]

First, let's look at the vertical position function. We will assume the second hand is six inches long and starts straight up in the "12 o'clock" position. At time 0, the point on the end of the hand will be as high above the horizontal-axis as possible (which is 6 inches). As the second hand moves clockwise, the point will go down for 30 seconds (reaching a low point of 6 inches below the horizontal axis), then go up for another 30 seconds, and repeat this process over and over. The points on the graph in Figure 11.4 represent the vertical distance from the horizontal axis of the tip of the second hand. Notice that the graph reflects the second hand's behavior of being straight up at 0 seconds and straight down at 30 seconds.

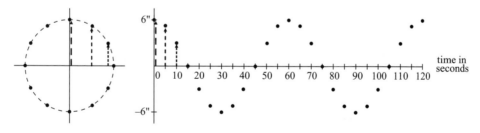

FIGURE 11.4
A graphical representation of the vertical position function.

To complete our graph, we connect the points with a smooth curve. (See Figure 11.5.) The result is a function that oscillates from a minimum height of −6 to a maximum height of +6 and repeats this process every 60 seconds.

Describing the horizontal position of the tip of the second hand on a clock is similar. As the hand starts on the 12 and moves clockwise, its horizontal position moves to the right. It will continue to move to the right until it reaches the 3 when the horizontal distance is +6. It will then move to the left and continue to do so past the 6 (the middle

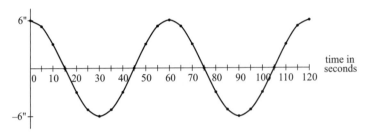

FIGURE 11.5
The points of the vertical position function connected with a smooth curve.

of the clock) until it reaches the 9 when the horizontal distance is −6. From the 9 to the 12, it will move to the right again.

The process of graphing the function representing the horizontal position is similar to graphing the vertical position. The input is still time (in seconds), but the output is now the horizontal distance to the right (or left) of the vertical axis. Notice that we are graphing the *horizontal* distance on the clock as *vertical* distance on our graph. (See Figure 11.6.) The graph of the horizontal distance is similar to the graph of the vertical distance, but has a different *y*-intercept. This is because the 12 o'clock position is 6 inches above the horizontal axis on the clock but 0 inches from the vertical axis on the clock.

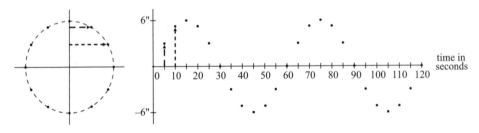

FIGURE 11.6
A graphical representation of the horizontal position function.

Both the vertical and horizontal position of the tip of a second hand represent periodic functions. The graphs repeat themselves every 60 seconds so the period of both the vertical position function and the horizontal position function is 60. In other words, for any two inputs that are 60 seconds apart, the outputs are the same.

Amplitude and Period

Two important terms associated with periodic functions are period and amplitude. Remember that the **period** is the *minimum* fixed distance between input values for which the outputs are the same. The period of a second hand on a clock is 60 seconds. This is the time it takes to complete one revolution. It is also the time (or distance) it takes the graph to complete one cycle. The start of this cycle can occur at any point on the graph.

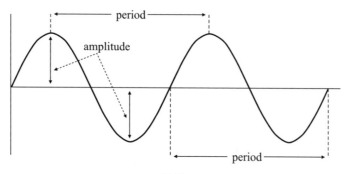

FIGURE 11.7
Amplitude and period.

Notice that, wherever it starts, 60 seconds later you have the same output. In business news, we often hear about seasonally adjusted data. This indicates that a cycle repeats every year, so the period is 12 months.

While the period is determined by the input of a function, the amplitude is determined by the output. If a horizontal line is drawn through the graph of a periodic function halfway between the function's maximum and the minimum values, the **amplitude** is the distance from this horizontal line to the maximum (or minimum) value. (See Figure 11.7.) The amplitude is defined symbolically as $\frac{1}{2}(M - m)$ where M is the maximum output and m is the minimum output. The amplitude of the second hand function is 6 inches. This is the length of the second hand itself and hence $M = 6$, $m = -6$, and amplitude $= \frac{1}{2}(6 - (-6)) = \frac{1}{2}(12) = 6$.

In the clock example, both the horizontal and vertical function had the same period and amplitude. This is not always the case. Suppose two-year-old Emma has a small swing in her back yard. As she swings, the functions that describe her position in an xy coordinate system are periodic.

A drawing of the dimensions of the path of the swing is shown in Figure 11.8. At the lowest, the swing is 0.5 meters above the ground while at the highest point, the swing

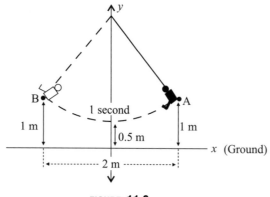

FIGURE 11.8
The swing function.

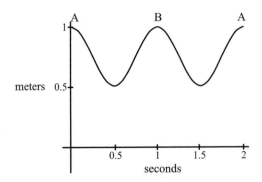

FIGURE 11.9
A graphical representation of the vertical position function of Emma's swing.

is 1 meter above the ground. The horizontal distance traveled is 2 meters, and the length of time it takes to swing from point A to point B is 1 second. An xy-coordinate system has been included in Figure 11.8. The x-axis is at ground level and the y-axis is where the swing would hang straight down.

Let's graph the vertical position for Emma's swing. Start at point A, the point farthest to the right. The first point on our graph is $(0, 1)$ since at the starting time, $t = 0$, Emma is 1 meter above the ground. Assume that at $t = 0.5$ seconds (halfway through the first swing), Emma is 0.5 meters above the ground. At $t = 1$ second, Emma is at point B which is 1 meter above the ground. She then starts to swing back towards point A which she reaches when $t = 2$ seconds. Figure 11.9 shows the graph of Emma's vertical position. Notice that the period is 1 second and the amplitude is $\frac{(1-0.5)}{2} = \frac{0.5}{2} = 0.25$ meters.

The horizontal position is also periodic but with a different period and amplitude. The horizontal position is measured from y, the vertical axis. At point A, Emma is 1 meter to the right of the vertical axis so our starting point on the graph is $(0, 1)$. Assume that at $t = 0.5$ seconds (halfway through the first swing), the swing is straight down so the point on the graph is $(0.5, 0)$. When $t = 1$ second, Emma is at point B. This is 1 meter *to the left* of the vertical axis so the point is $(1, -1)$. Emma then starts back towards point A which she reaches when $t = 2$. Figure 11.10 shows the graph of Emma's horizontal position. Notice that the period is 2 seconds and the amplitude is $\frac{1-(-1)}{2} = \frac{2}{2} = 1$ meter.

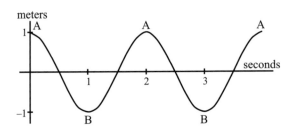

FIGURE 11.10
A graphical representation of the horizontal position function of Emma's swing.

Graphs of both the vertical and horizontal position of Emma's swing represent periodic functions. They are similar to those in the clock example in many respects. All four graphs have the same basic shape. However, both of the clock graphs are centered at the x-axis while the vertical swing position graph is centered at $y = 0.75$. In the clock example, both the horizontal and vertical position functions have the same amplitude and period while in the swing example the amplitude and period of the two functions are different.

Sine and Cosine Functions

The sine and cosine functions are used to describe many types of periodic behavior. You may be familiar with the right triangle definitions of these functions:

$$\sin\theta = \frac{\text{opposite}}{\text{hypotenuse}} \quad \text{and} \quad \cos\theta = \frac{\text{adjacent}}{\text{hypotenuse}}.$$

Another way to define the sine and cosine functions is to use the vertical and horizontal position of a point on a unit circle.[3] Defining sine and cosine in terms of position on a circle is similar to looking at the vertical and horizontal position functions of the second hand of a clock. This implies that the sine and cosine functions are periodic.

Consider an arbitrary point, P, on a unit circle that is centered at the origin as shown in Figure 11.11.

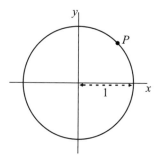

FIGURE 11.11
A unit circle with an arbitrary point P.

The sine and cosine functions are defined in terms of the vertical and horizontal position of point P. The input is the distance on the circle from $(1, 0)$ *counter-clockwise* to P. In other words, x is the arc length from $(1, 0)$ to the point P. **Sine** of x (abbreviated $\sin x$) is the vertical position of P and **cosine** of x (abbreviated $\cos x$) is the horizontal position of P. The coordinates of P are therefore $(\cos x, \sin x)$. (See Figure 11.12.)

One complete revolution around the circle has length 2π.[4] Because the input is the arc length counter-clockwise from $(1, 0)$, inputs for sine and cosine are often given in terms

---

[3] The radius of a unit circle is 1.

[4] Remember that the formula for the circumference of a circle is $C = 2\pi r$ and, since our radius is 1, the circumference of this circle is 2π.

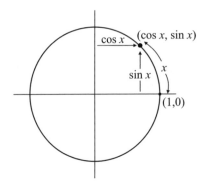

FIGURE 11.12
A unit circle showing the point (cos x,sin x).

of π. Halfway around the circle is π, one quarter of the way is $\frac{\pi}{2}$, and so on. Inputs larger than 2π mean you have gone around the circle more than one revolution. Negative inputs represent arc lengths which proceed clockwise from $(1, 0)$ rather than counter-clockwise. Since the coordinates of the points on the circle repeat after we have gone around the circle an arc length of 2π, the period of the sine and cosine functions is 2π.

Constructing graphs of the sine and cosine functions is similar to constructing graphs for the vertical and horizontal clock functions. Instead of time as the input, these functions have arc length as the input. We start these functions at the point $(1, 0)$ (the 3 o'clock position on a clock), and proceed counterclockwise rather than clockwise. Since arc length is the input, it makes sense to have the circumference of the unit circle transformed into the x-axis for our graph. Think of this as cutting the circle at $(1, 0)$ and unrolling. (See Figure 11.13.)

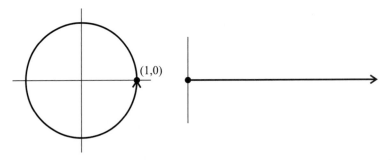

FIGURE 11.13
A unit circle transformed into the x-axis.

To graph $y = \sin x$, remember that $\sin x$ is defined as the y-coordinate (vertical distance) of point P on the unit circle. Thus, the graph of $y = \sin x$ tells us how far P is above (or below) the x-axis. To show this, we will plot eight equally-spaced points around the unit circle. This subdivides the circumference of the circle, 2π, into eight

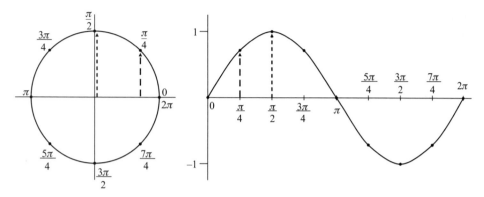

FIGURE **11.14**
The graphical form of $y = \sin x$.

equal lengths of $\frac{\pi}{4}$. We will then transfer the height of these points to our graph and connect the points with a smooth curve as shown in Figure 11.14.

Determining the graph of $y = \cos x$ is similar. We will again plot eight points around the unit circle. We will then transfer the horizontal distance of each of these points from the y-axis to our graph and connect the points with a smooth curve. (See Figure 11.15.) The amplitude of the $\cos x$ function is 1 since $\frac{1}{2}(1 - (-1)) = 1$. The amplitude of the $\sin x$ function is also 1 for the same reason. The period of both functions is 2π.

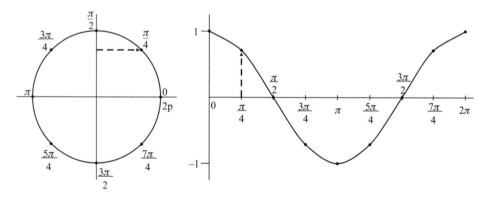

FIGURE **11.15**
The graphical form of $y = \cos x$.

Notice that we are using arc length, rather than an angle, as the input for our sine and cosine functions. You are probably accustomed to using an angle as the input. Using **radians**, rather than degrees, to measure the angle allows us to equate input as arc length with input as an angle. A radian is defined as the angle (measured counterclockwise from the positive x-axis) that "sweeps out" an arc length of one unit on the unit circle. Going all the way around the unit circle one time is 2π radians since the arc length is 2π.

Notice that all our graphs of cyclic behavior are quite similar. In fact, the graph of $y = \sin x$ is itself similar to the graph of $y = \cos x$. In some sense, they are the same graph except that one starts at $(0, 0)$ and the other one starts at $(0, 1)$. The shape and behavior of these two functions, $y = \sin x$ and $y = \cos x$, is what makes them the appropriate choice for modeling cyclic phenomenon.

Application: Ferris Wheel

In July 1997, a huge Ferris wheel opened up at an amusement park in Osaka, Japan. The wheel is 112.5 meters from the ground to the top and has a diameter of 100 meters. It holds 60 cabins that seat up to eight people each. One revolution of the wheel takes 15 minutes.[5] Suppose you get on the Osaka Ferris wheel at its lowest point (12.5 meters above the ground) and it turns clockwise for two revolutions. The graph in Figure 11.16 shows your height above the ground on the vertical axis and the minutes after the wheel begins to turn on the horizontal axis. Note that the amplitude is 50 meters and the period is 15 minutes.

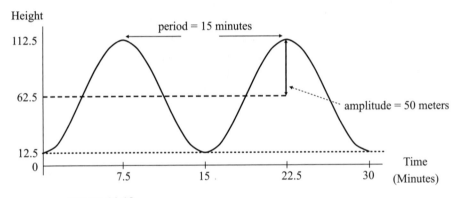

FIGURE 11.16
A graph of the height of a person riding on the Osaka Ferris wheel.

Example 1. Suppose you get on the Osaka Ferris wheel at its lowest point and it turns clockwise for two revolutions. This time, however, assume it takes the wheel 20 min to complete one revolution. Construct a graph with your height above the ground on the vertical axis and the minutes after the wheel begins to turn on the horizontal axis. Label the period and amplitude on your graph. Assume the wheel revolves continuously and goes slowly enough so that you can hop on and off.

Solution. If the time to complete one revolution is now 20 minutes, the period of the new graph is 20 minutes. Notice that the amplitude, however, is the same as the original graph. The graph is shown in Figure 11.17. ∎

Any situation that repeats itself in the same pattern (such as the path of a car on a Ferris wheel) can be modeled with a periodic function.

---

5 Japan Information Network, "World's Tallest Ferris Wheel in Osaka," *Monthly News*, August 1997 (visited 4 November 1997) <www.jinjapan.org/kidsweb/news/97-8/wheel.html>.

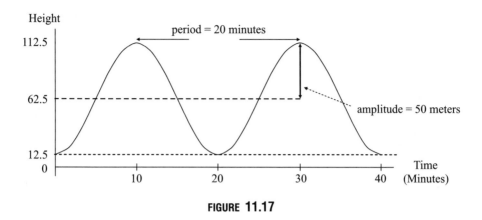

FIGURE 11.17

Reading Questions for Periodic Functions

1. Why can't you use an exponential function to model periodic behavior?

2. Why is graph (d) in Figure 11.2 not a function?

3. Give an example of something (not in the reading) that can be described using periodic functions.

4. The graph of $y = f(x)$ is given below.

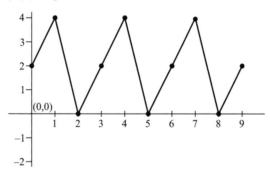

   (a) What is the period of $y = f(x)$?

   (b) Find the values for each of the following:

   i. $f(10)$ iii. $f(16)$
   ii. $f(11)$ iv. $f(18)$

5. Suppose $y = f(x)$ is a periodic function whose period is 3. If $f(0) = 1$, what is $f(6)$?

6. Let $y = S(d)$ be the function whose input is the day of the year and whose output is the total amount of possible daylight in your town. Explain why $y = S(d)$ is a periodic function. What is its period?

7. The shortest day of the year occurs at the winter solstice on approximately December 21 and the longest day is on the summer solstice on approximately June 21. For a

certain location, the shortest day has 9 hours of daylight and the longest day has 15 hours of daylight. Sketch a graph of the number of hours of daylight versus the day of the year over a two year period. On which days are there 12 hours of daylight?

8. Determine if each of the following graphs represents a periodic function.

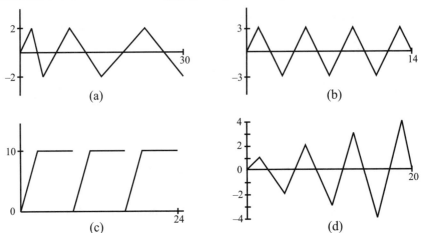

9. How would the graph of the vertical clock function change if the second hand was 4 inches long instead of 6 inches long?

10. Explain why the horizontal position function of the swing and the vertical position function have different periods.

11. What scenario would cause the amplitude of the horizontal position function for Emma's swing to change?

12. In March of 2000, the world's tallest Ferris wheel began operating at an amusement park in London, England. It is 135 meters from the ground to the top, and the wheel has a diameter of 120 meters. There are 32 cabins that seat up to 25 people each. One revolution of the wheel takes 30 minutes.[6] Suppose you get on the London Ferris wheel at its lowest point and you ride it clockwise for two revolutions. Sketch a graph with your height above the ground on the vertical axis and the minutes after the wheel begins to turn on the horizontal axis. Assume the wheel revolves continuously and goes slow enough so you can hop on and off. What is the period and the amplitude of your graph?

13. Suppose a man is pacing up and down a hallway. It is 8 feet from one end of the hall to the other and he paces this distance every 4 seconds. Let $d(t)$ be the man's distance from one end of the hall at time t.

   (a) Why is this function periodic?

   (b) What is the period of this function?

   (c) What is the amplitude?

---

[6] *British Airways London Eye*, 28 October 2002,
<http://www.geocities.com/coaster_feature1/londoneye.html>.

14. What is a unit circle?

15. What is the definition of $\sin x$ in terms of the unit circle?

16. What is the definition of $\cos x$ in terms of the unit circle?

17. Use the following figure to approximate the values of the sine and cosine functions. Each letter represents the length of the arc beginning at $(1,0)$ and ending at the indicated point.

   (a) $\cos A$
   (b) $\sin B$
   (c) $\sin C$
   (d) $\sin D$
   (e) $\cos E$
   (f) $\cos F$

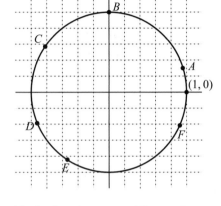

18. Match the graphs with the functions: $y = 2^x$, $y = 2x$, $y = \log x$, and $y = \sin x$.

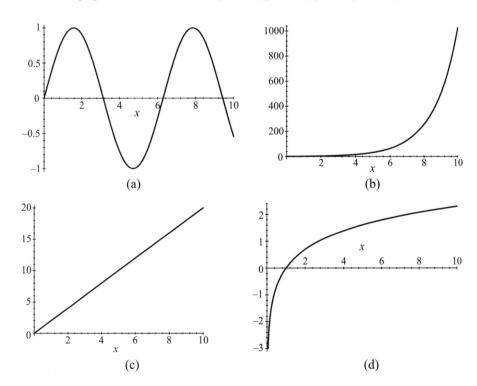

19. The child in the swing in the following figure takes 1.2 seconds to swing forward and 1.2 seconds to swing back. The maximum height the child is off the ground is 4

feet and the minimum height is 1.5 feet. The horizontal distance from the rightmost position to the leftmost position is ten feet.

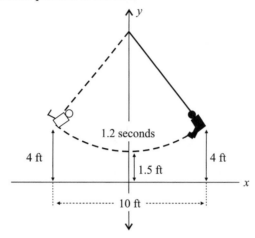

(a) Construct a graph of the vertical position of the swing as the child swings for 8 seconds. What are the period and amplitude?

(b) Construct a graph of the horizontal position of the swing as the child swings for 8 seconds. What are the period and amplitude?

20. Suppose they replace the Osaka Ferris wheel with a new one so that its highest point is 150 meters from the ground and the wheel has a diameter of 140 meters. (See Example 1.) It still completes one revolution in 15 min. You get on this new wheel at its lowest point and it turns clockwise for two revolutions. Construct a graph with your height above the ground on the vertical axis and the minutes after the wheel begins to turn on the horizontal axis for this situation. Label the period and amplitude on your graph. Assume the wheel revolves continuously and goes slow enough so that you can hop on and off.

Periodic Functions: Activities and Class Exercises

1. **Sound.**[7] Sound waves are an example of a periodic phenomenon that can be modeled with a sine or a cosine function. This activity uses the CBL and a microphone probe to explore some of the properties of sound waves.

Directions for the Sound Program

• You must first load the **Sound** program onto your calculator from your instructor's calculator. Only one member of your group needs to have this program.

------

[7] This activity requires the program SOUND.83p for the TI-83. It can be obtained on the Internet at <http://www.math.hope.edu/swanson/calculator/programs.html>.

- Plug the microphone probe into Channel One (top of the CBL) and connect the CBL to your calculator (using the short cord in the CBL box).

- Choose the **Sound** program and press $\boxed{\text{ENTER}}$ until you get to the screen that says "Press ENTER to collect data."

- BEFORE starting to collect data, hit the tuning fork with the rubber mallet. When you have a sustained note, put the tuning fork close to the microphone and begin collecting data.

- When the CBL is finished collecting data, use $\boxed{\text{TRACE}}$ to find the x-coordinates at the beginning and end of one period of the graph. Subtract the x-values to find the period. DO NOT ROUND.

- To do another graph, press $\boxed{\text{ENTER}}$ and/or $\boxed{\text{Clear}}$ until you are back at the starting screen for the program.

(a) Using five different tuning forks, complete the table below. The frequency, or the number of vibrations per second, is listed on the tuning fork. You can determine the x-coordinates by using the `Trace` button on your calculator. The period of the sound wave is the difference in x-coordinates. The last column is asking for the reciprocal of the period. You should notice that the numbers in the second column (Frequency) are similar to the numbers in the last column (1/(Period of Sound Wave)). This is because $f = \frac{1}{P}$ where f is the frequency of the note and P is the period of the sound wave.

Note	Frequency	x-coordinate at peak	x-coordinate at adjacent peak	Period of Sound Wave	1/(Period of Sound Wave)

(b) i. Get a glass bottle with a narrow opening at the top. Blow across the lip of the bottle to produce a sustained note. Use the Calculator Based Laboratory (CBL) to record the period of the sound wave. Fill the bottle partially with water and repeat this procedure TWICE. In total, you should have produced three different notes and completed the following table.

x-coordinate at peak	x-coordinate at adjacent peak	Period of Sound Wave	1/(Period of Sound Wave)

ii. We are now interested in using the sound waves to determine what note was produced. Below is a chart of various notes and their respective frequencies. If your frequency does not appear in the chart, use the fact that a note with the same name which is an octave lower is $1/2$ the frequency. For example, the C which is one octave lower than the first C listed in the table has a frequency of $\frac{262}{2} = 131$. What three notes did you produce?

Note	C	C# or Db	D	D# or Eb	E	F	F# or Gb
Frequency in Hz	262	277	294	311	330	349	370

Note	G	G# or Ab	A	A# or Bb	B	C (next octave)
Frequency in Hz	392	415	440	466	494	524

(c) Now that you have used tuning forks and bottles to produce notes, we are interested in seeing sound waves produced by your voice. Do each of the following and record your observations.

   i. Have one person hum a note. Describe (or sketch) the sound wave. What note did the person hum?

   ii. Have a number of people hum the same note. Describe (or sketch) the sound wave. Notice any differences? [Note: Do not find the period.]

   iii. Have two people hum the same note an octave apart. What happens this time? [Note: Do not find the period.]

   iv. Have two or three people hum in harmony. Now what do you notice? Why do you think this is happening? [Note: Do not find the period.]

2. **Biorhythms.** Some people believe that our behavior is partially governed by biorhythms. Typically, biorhythms consist of three cycles (physical, emotional, and intellectual) which begin at birth. According to proponents of this theory, when our cycles are positive we feel better physically, are more cheerful, and think more quickly. The physical cycle has a period of 23 days and a formula of $P = \sin(\frac{2\pi t}{23})$; the emotional cycle has a period of 28 days and a formula of $E = \sin(\frac{2\pi t}{28})$; and the intellectual cycle has a period of 33 days and a formula of $I = \sin(\frac{2\pi t}{33})$. In all these formulas, the input, t, is the number of days since birth.

(a) To use the biorhythm functions, you must first compute your age in terms of days.

   • Multiply your age by 365.

   • Determine the number of leap years since you were born. They occur in years that are multiples of 4. For example, 1980, 1984, 1988, and so on are leap years. If you were born during a leap year, only include it if you were born in January or February. If the current year is a leap year, only include it if today is after February 28.

- Determine the number of days since your last birthday. [Hint: April, June, September, and November have 30 days, February has 28 days, and all the rest have 31 days.]

- Find the sum of your numbers from the three steps above. This is your age in days.

(b) Use the formulas to find your current value for each of the 3 biorhythms. When doing this, make sure your calculator is in radian mode. Do these values seem to correspond to the way you are feeling physically, emotionally, and intellectually today? Remember that positive numbers are good and negative numbers are bad.

(c) Graph the three biorhythms functions for the next two weeks. This means you will graph $P = \sin(\frac{2\pi t}{23})$, $E = \sin(\frac{2\pi t}{28})$, and $I = \sin(\frac{2\pi t}{33})$. Use an appropriate domain. For example, if you are 7000 days old, use a domain of 7000 to 7014. Sketch your graphs, being sure that each curve is clearly labelled. Include points on each graph corresponding to today as well as points corresponding to the day of your next test. Make sure these points are labeled clearly.

(d) Do you think there is validity to the theory of biorhythms? Why or why not?

3. **Alternating Current.**[8] Electric service is provided in the form of *alternating* current (AC) at 110 volts and 60 hertz. This means that the voltage cycles between -110 volts and $+110$ volts, and that 60 cycles occur each second. This activity allows you to see the periodic behavior of alternating current using the Calculator Based Laboratory (CBL) and the light probe. Answer the following questions using the graph you obtained from the CBL.

Directions for using the light probe

- You must first load the **Physics** program onto your calculator. Only one member of your group needs to have this program. Be sure to load ALL the programs that begin with PHZ as well. [Note: This program is fairly large. You may have to delete the other programs on your calculator in order to have enough memory.

- Plug the light intensity probe into Channel One (top of the CBL) and connect the CBL to your calculator (using the short cord in the CBL box).

- Choose the **Physics** program. Press ENTER until you get to the **Main Menu**.

- Choose **1: Set up Probes**. The number of probes is 1. Select **More probes** and the select **Light**. The channel number is 1.

---

[8] This activity requires the program PHYSICS.83g for the TI-83. It can be obtained on the Internet at <http://www.math.hope.edu/swanson/calculator/programs.html>.

- You should now be back at the **Main Menu**. This time, choose **2: Collect Data**. Next, choose **2: Time Graph**.

- Use 0.0005 as the time between samples and use 200 as the number of samples.

- When it says **Continue?**, choose **1: Use Time Setup**. Hold sensor up to a fluorescent light. Press ENTER .

- When the data collection is done, press ENTER to see the graph.

- Use TRACE to find the period of the graph. DO NOT ROUND. When you are done, press ENTER , **2:NO**, and **6:Quit**. You can still see your graph if you press GRAPH .

(a) What do the peaks or maximum values represent in terms of the fluorescent bulb and the current? What do the minimum values represent?

(b) Find the period of your graph.

(c) The period represents the time required for one complete on-off cycle; that is, the seconds per cycle. Find the frequency (cycles per second) by finding $\frac{1}{p}$ where p is the period.

(d) How is the frequency of your graph related to the fact that the current frequency is 60 cycles per second?

4. **Daylight.** The function whose input is a day of the year and whose output is the maximum possible hours of daylight on that day is periodic. We are interested in determining mathematical properties of this function and exploring how they are related to the seasons.

   Below is a table giving the minutes of daylight for every 10th day (where Day 1 is January 1st) for Grand Rapids, Michigan.[9]

   (a) Make a scatterplot using the points from the table with the input representing the day of the year (numbered 1 through 365) and the output representing the number of minutes of daylight.

   (b) Why is this function periodic?

   (c) Which day had the most daylight? What date is this? Which day had the least daylight? What date is this? Does this correspond with your experience and your knowledge of the seasons? Explain.

   (d) Looking at the graph, find the time of year when the amount of daylight is changing most rapidly. When is it changing the least? How does this correspond with your experience and your knowledge of the seasons? Explain.

----

[9] Sunrise and sunset times for any location worldwide can be found at
<http://aa.usno.navy.mil/data/docs/RS_OneYear.html>.

Day	Hrs	Min	Total Minutes
1	9	5	545
11	9	16	556
21	9	33	573
31	9	55	595
41	10	20	620
51	10	47	647
61	11	13	673
71	11	42	702
81	12	11	731
91	12	40	760
101	13	9	789
111	13	36	816
121	14	3	843
131	14	27	867
141	14	49	889
151	15	5	905
161	15	16	916
171	15	21	921
181	15	19	919

Day	Hrs	Min	Total Minutes
191	15	11	911
201	14	56	896
211	14	37	877
221	14	14	854
231	13	49	829
241	13	23	803
251	12	54	774
261	12	26	746
271	11	57	717
281	11	28	688
291	10	59	659
301	10	32	632
311	10	6	606
321	9	43	583
331	9	24	564
341	9	9	549
351	9	2	542
361	9	2	542

(e) What is the amplitude of your function?

(f) What is the period of your function?

12

Power Functions

We have looked at linear, exponential, logarithmic, and periodic functions in previous sections. The last type of function we'll consider is a power function. Power functions are used to describe things such as an accelerating car or the area of a circle. In this section, we will explore characteristics of power functions with positive exponents and examine how to distinguish them from other types of functions.

Definitions and Examples

Suppose you are ordering a pizza at a restaurant. You notice that an 8-inch personal pizza is supposed to feed one person while the large 16-inch pizza is supposed to feed four. You might find it interesting that doubling the diameter of the pizza results in four times as many people being able to eat it. Using the formula for the area of a circle,

$$A = \pi r^2,$$

you can easily find the area of each of these pizzas.

Since the size of a pizza is typically given in terms of its diameter and our area formula uses radius as the input, we need to divide each of the two diameters by two. This gives us

$$A = \pi \cdot 4^2 = 16\pi \approx 50.3 \text{ sq. in.} \quad \text{and} \quad A = \pi \cdot 8^2 = 64\pi \approx 201.2 \text{ sq. in.}$$

This shows that the 16-inch pizza is four times as large as the 8-inch personal pizza.

A graph of this area function, $A = \pi r^2$, is shown in Figure 12.1. It is obviously not a linear function since it does not look like a straight line. While it does look somewhat like an exponential function, it is not since the variable in the formula is not in the exponent and the graph goes through the origin.[1] This is an example of a power function.

---

[1] Recall that an exponential function is of the form $y = ba^x$ where a is the growth factor and b is the y-intercept.

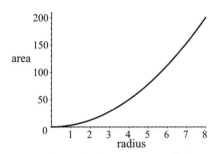

FIGURE 12.1

The area of a circular pizza, in square inches, given the length of its radius, in inches.

A **power function** is a function of the form $f(x) = kx^p$ where k and p are constants. Note that this looks similar to an exponential function. However, with an exponential function the *base* remains constant while with a power function the *exponent* (or power) remains constant.

You may already be familiar with some power functions such as $f(x) = x^2$ and $g(x) = x^3$. The graphs of these two functions are shown in Figure 12.2.

The graphs of $f(x) = x^2$ and $g(x) = x^3$ look very similar for positive inputs (in the first quadrant).[2] For negative inputs, however, they look quite different since $g(x) = x^3$ gives negative outputs and $f(x) = x^2$ does not. Since most applications of power functions do not include negative inputs, we will only look at inputs greater than 0.

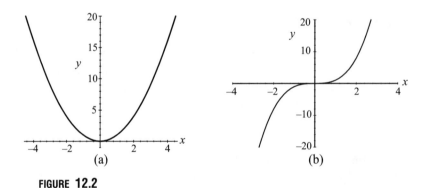

FIGURE 12.2

Two basic power function shapes. (a) is $f(x) = x^2$ and (b) is $g(x) = x^3$.

Properties of Exponents

Before looking at the effect the exponent or power, p, has on power functions, let's briefly review some properties of exponents. Recall that an integer exponent is a convenient way of denoting multiplication of a number by itself. For example, $2^3 = 2 \cdot 2 \cdot 2 = 8$. The exponent or power indicates the number of twos we need to multiply together.

---

[2] The first quadrant is the upper right-hand portion of the graph where x and y are both positive.

An exponent of one simply means you have one number (so there is nothing to multiply). So $2^1 = 2$. There is another integer exponent, namely zero, which may seem strange. If your exponent is zero, then you do not have any copies of your number to multiply. For consistency in properties of exponents, we define any nonzero number raised to an exponent of zero to be equal to 1. So $2^0 = 1$, $1^0 = 1$, and $100^0 = 1$. Any nonzero number raised to an exponent of zero is one *by definition*.

Sometimes the exponent is not an integer, but is a rational number instead. A **rational number** is one that can be written as the quotient of two integers. In other words, a rational number is a fraction. So what does it mean to have something like $2^{1/3}$? Obviously, you cannot multiply two by itself one-third times. To remedy this quandary, we define $x^{1/n} = \sqrt[n]{x}$. If an exponent is written as a fraction, the denominator indicates taking a *root* of that number. So

$$25^{1/2} = \sqrt{25} = 5 \quad \text{and} \quad 2^{1/3} = \sqrt[3]{2} \approx 1.26.$$

The calculator screen for evaluating $2^{1/3}$ is shown in Figure 12.3. The parentheses in the calculator commands are essential. Without them, the calculator interprets $2 \wedge 1/3$ as $\frac{2^1}{3}$ which is **not** what you want!

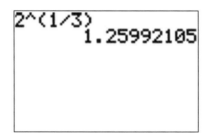

FIGURE 12.3

A calculator screen giving a decimal approximation to $2^{(1/3)}$.

Let's look at the meaning of the exponent when you have a fraction where the numerator is not one. In this case, you combine the two ideas. Think of the numerator of the fractional exponent as the number of copies of that number you want to multiply by itself and think of the denominator as a root. For example,

$$2^{2/3} = (2^{1/3})^2 = (\sqrt[3]{2})(\sqrt[3]{2}) = \sqrt[3]{2 \cdot 2} = \sqrt[3]{4} \approx 1.59.$$

Example 1. Write each of the following using radicals (i.e., write $x^{1/2}$ as $\sqrt{x}$) and simplify.

1. $4^{(1/2)}$

2. $4^{(5/2)}$

3. $27^{(1/3)}$

4. $27^{(3/2)}$

Solution.

1. $4^{(1/2)} = \sqrt{4} = 2$
2. $4^{(5/2)} = (\sqrt{4})^5 = (\sqrt{4})(\sqrt{4})(\sqrt{4})(\sqrt{4})(\sqrt{4}) = \sqrt{4 \cdot 4 \cdot 4 \cdot 4 \cdot 4} = \sqrt{1024} = 32$
3. $27^{(1/3)} = \sqrt[3]{27} = 3$
4. $27^{(3/2)} = (\sqrt{27})^3 = (\sqrt{27})(\sqrt{27})(\sqrt{27}) = \sqrt{27 \cdot 27 \cdot 27} \approx 140.30$ ■

The Effect of the Value of p

With exponential functions, $y = ba^x$, the value of a was crucial in determining the shape of the graph. For power functions, $y = kx^p$, it is the value of p that determines the shape of the graph. Let's compare three cases: p greater than 1, p between 0 and 1, and p less than 1.[3]

Power functions with $p > 1$

We will start by exploring the connections between the equation, the shape of the graph, and the behavior of the data for functions of the form $f(x) = x^p$ where p is positive. Values for three different power functions and their graphs on a domain of $0 \leq x \leq 4$ are shown in Figure 12.4.

What can be concluded from these data? We see that each of the three functions, $y = x^2, y = x^3, y = x^4$, contain the points $(0, 0)$ and $(1, 1)$ and that each function is increasing. We also see that, when $x > 1$, the function increases faster when the exponent or power is larger. Looking at the graphs, it is obvious that each of these functions is concave up in the first quadrant. Recall that concave up means the rate of increase is increasing. In this case, that means the graphs are getting steeper for larger values of x.

x	$y = x^2$	$y = x^3$	$y = x^4$
0	0	0	0
1	1	1	1
2	4	8	16
3	9	27	81
4	16	64	256

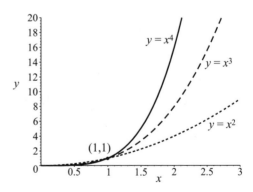

FIGURE 12.4

Comparing power functions where the power, p, is greater than 1.

Power functions with $0 < p < 1$

We just looked at power functions with exponents or powers greater than 1. Now let's look at the behavior when $0 < p < 1$. We are again interested in making connections

---

[3] If p is either 0 or 1, the power function becomes a linear function, so we will not consider these cases.

x	$y = x^{1/3}$	$y = x^{1/2}$	$y = x^{2/3}$
0	0	0	0
1	1	1	1
2	≈ 1.26	≈ 1.41	≈ 1.59
3	≈ 1.44	≈ 1.73	≈ 2.08
4	≈ 1.58	2	≈ 2.52

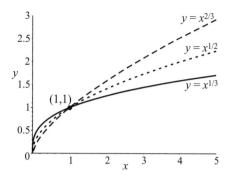

FIGURE 12.5

Comparing power functions when $0 < p < 1$.

between the symbolic formula, the shape of the graph, and variations in the data. Values (rounded to two decimal places) for three power functions and their graphs on a domain of $0 \le x \le 4$ are shown in Figure 12.5.

Once again, all of these functions ($y = x^{1/3}, y = x^{1/2}, y = x^{2/3}$) contain the points $(0, 0)$ and $(1, 1)$ and are increasing. It is also true that, when $x > 1$, the function increases faster when the exponent is larger. However, the rate at which these functions are increasing is slowing down as x gets larger. For example, looking at $y = x^{1/3}$, you can see that it increases by approximately 0.26 units between $x = 1$ and $x = 2$, and increases approximately 0.14 units between $x = 3$ and $x = 4$. From the graphs in Figure 12.5 you can see that when the rate of increase is decreasing, the result is graphs that are concave down.

So far, we have seen that power functions with $p > 1$ are concave up in the first quadrant and those with $0 < p < 1$ are concave down. This should be somewhat intuitive since the linear function $y = x$ can be thought of as the power function where $p = 1$ (i.e., $y = x^1$) and those with exponents greater than 1 should increase at a faster rate (concave up) while those with exponents less than 1 should increase at a slower rate (concave down).

Power functions with $p < 0$

Now that we have considered power functions with $p > 1$, and $0 < p < 1$, we are left to consider power functions where $p < 0$. We are still interested in making connections between the symbolic formula, the shape of the graph, and variations in the data. Values (rounded to two decimal places) for three power functions and their graphs on a domain of $0 \le x \le 4$ are shown in Figure 12.6.

Once again, all of these functions ($y = x^{-1/2}, y = x^{-1}, y = x^{-2}$) have the point $(1, 1)$ in common. However, this time they do not contain the point $(0, 0)$. In fact, these functions are all undefined when $x = 0$. This is because a negative exponent or power results in taking the reciprocal. In other words, the negative input causes the input to become the denominator of a fraction. For example, $y = x^{-1}$ can be rewritten as $y = \frac{1}{x}$. Since the input is in the denominator and since you cannot divide by 0, power functions with $p < 0$ are undefined for $x = 0$.

x	$y = x^{-1/2}$	$y = x^{-1}$	$y = x^{-2}$
0	undefined	undefined	undefined
1	1	1	1
2	≈ 0.71	0.5	0.25
3	≈ 0.58	≈ 0.33	≈ 0.11
4	0.5	0.25	≈ 0.06

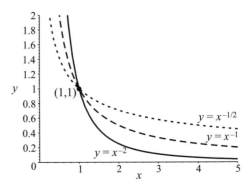

FIGURE 12.6

Comparing power functions where $p < 0$.

The other noticeable difference is that these functions are decreasing as x gets larger. So while power functions where $p > 0$ are increasing in the first quadrant, power functions where $p < 0$ are decreasing in the first quadrant. Looking at the graphs of these functions in Figure 12.6, you can see that they are concave up. Concave graphs have the property that the rate is increasing. However, since the rate is negative (i.e., the graphs are decreasing), this means that the rate at which these functions are decreasing is slowing down.

Summary of Properties

The following is a summary of the properties of $f(x) = x^p$ when $x \geq 0$:

- Points in common
    * $(0, 0)$ if $p > 0$.
    * $(1, 1)$ for all power functions.
- Shape and Direction
    * If $p > 1$, the function is increasing and concave up in the first quadrant.
    * If $0 < p < 1$, the function is increasing and concave down in the first quadrant.
    * If $p < 0$, the function is decreasing and concave up in the first quadrant.
- For $p > 0$, the larger the power, the faster the function is increasing and the steeper the graph for $x > 1$.
- For $p < 0$, the smaller the power (i.e., the more negative), the faster the function is decreasing and the steeper the graph for $0 < x < 1$.

Other Values for k

So far, we have considered power functions of the form $f(x) = x^p$ and concentrated on the value of p. However, a general power function is of the form $f(x) = kx^p$ where both k and p are constants. What happens if $k \neq 1$? Let's start to answer this question by comparing $y = 0.5x^2$, $y = x^2$, and $y = 4x^2$. Values for these three power functions and their graphs on a domain of $0 \leq x \leq 3$ are shown in Figure 12.7.

x	$y = 0.5x^2$	$y = x^2$	$y = 4x^2$
0	0	0	0
1	0.5	1	4
2	2	4	16
3	4.5	9	36

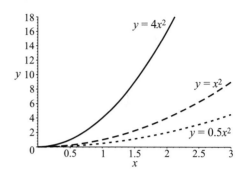

FIGURE 12.7

Comparing $y = x^2$, $y = \frac{1}{2}x^2$, and $y = 4x^2$.

All of the functions, $y = 0.5x^2, y = x^2, y = 4x^2$, are increasing and concave up, although $y = 4x^2$ is steeper than $y = x^2$ while $y = 0.5x^2$ is not quite as steep. Notice what happens when $x = 1$. Earlier in this section, we observed that all power functions of the form $y = x^p$ contain the point $(1, 1)$. That is no longer true for power functions of the form $y = kx^p$ where $k \neq 1$. The function $y = x^2$ still contains the point $(1, 1)$ but $y = 0.5x^2$ contains the point $(1, 0.5)$ while $y = 4x^2$ contains the point $(1, 4)$. However, the behavior of the functions at $x = 0$ has not changed. All three of the functions still go through the point $(0, 0)$.

Let's investigate why $f(x) = kx^p$ for $k \neq 1$ changes the behavior of the power function for $x = 1$ but not for $x = 0$. The answer is quite simple—just evaluate $f(x) = kx^p$ for both $x = 1$ and $x = 0$. We have $f(1) = k \cdot 1^p = k \cdot 1 = k$ since 1 raised to any power is still 1. So power functions of the form $f(x) = kx^p$ contain the point $(1, k)$. This makes it easy to find the value of k from a table of numbers or to estimate it from a graph—just look at the output for $x = 1$. To see what happens when $x = 0$, notice that $f(0) = k \cdot 0^p = k \cdot 0 = 0$ as long as $p > 0$. So *every* power function contains the point $(0, 0)$ regardless of the value of k, as long as $p > 0$.

Other properties of power functions are not impacted by the value of k (as long as k is positive). What was mentioned earlier with regards to direction and concavity still holds true. The important difference in using a value of k other than 1 is that:

- Points in common for $f(x) = kx^p$ are:
  * $(0, 0)$ if $p > 0$.
  * $(1, k)$ for all values of p.

Distinguishing Power Functions from Other Functions

We have looked at linear, exponential, logarithmic, periodic, and power functions. While power functions are easily distinguished from some of these types (such as linear), they exhibit similar properties to other types (such as exponential). How can power functions easily be distinguished from the other types of functions?

Let's start by considering linear and periodic functions. Linear functions have a constant slope and appear as lines in a graph. No other function has these characteristics. Periodic functions oscillate up and down in some sort of fashion. No other function in the five types we have studied has these characteristics. So it is easy to distinguish power functions from either linear or periodic functions.

This leaves us with exponential and logarithmic functions. Exponential functions can be increasing and concave up or decreasing and concave up. Power functions can also be both of these. Logarithmic functions are always increasing and concave down. So are some power functions. However, remember that power functions always go through the point $(0,0)$ if $p > 0$ and are undefined at $x = 0$ if $p < 0$. An exponential function or a logarithmic function that is not shifted will never go through the point $(0,0)$. Exponential functions will never have an output that is 0 for any value of x while logarithmic functions are undefined for $x = 0$. Logarithmic functions are concave down and power functions for $p < 0$ are concave up. So even though both of these are undefined for $x = 0$, their graphs have different shapes.

Power Regression

Given just a couple of data points, we found formulas for linear and exponential functions in previous sections. We did this quite easily because we could use the y-intercept and either the constant slope or the constant growth factor to find the equation. When given data representing a power function of the form $f(x) = kx^p$, we can easily find the value of k since it contains the point $(1, k)$, but finding the value of p is more difficult. We do not have a constant slope or a constant growth factor to help us out. Using another point and algebra can also be tricky. Therefore, to find the equation of a power function from data, we are going to use power regression.

Doing power regression on a graphing calculator consists of almost the same process as linear regression. Typically, the data is entered in two lists and a power regression function is used to find a power function that fits the data. Similar to linear regression, if the data perfectly fits a power function, the calculator will give you the correct solution and the correlation will be 1 or -1. However, if the data do not perfectly fit a power equation, power regression will give a function that fits fairly well.[4]

Example 2.   Determine formulas for each of the following functions in Table 12.1. One is linear, one is exponential, and one is a power function.

TABLE **12.1**

x	0	1	2	3	4
$f(x)$	-5	-3	-1	1	3
$g(x)$	4	8	16	32	64
$h(x)$	0	3	12	27	48

---

[4] Power regression does not always minimize the sum of the errors between the output of the data and the regression equation like linear regression does.

Solution.   A linear function has a constant rate of change. The first function, f, is linear because as each input increases by 1, each output increases by 2. This means the rate of change, or slope, is 2. The y-intercept for this function is -5 since that is the output when the input is 0. Therefore the formula for this function is $f(x) = 2x - 5$.

An exponential function has a constant growth factor. The second function, g, is exponential because as each input increases by 1, each output increases by a factor of 2. This means the growth factor is 2. The y-intercept for this function is 4 since that is the output when the input is 0. Therefore the formula for this function is $g(x) = 4 \cdot 2^x$.

Since f is linear and g is exponential, h must be a power function. As with all power functions where $p > 0$, we can see that it contains the point $(0, 0)$. We can also see that for a power function of the form $y = kx^p$, k must be 3 since this is the output when the input is 1. This means our formula is $h(x) = 3x^p$. Using power regression on a calculator we find that the value of p is 2 and our formula is $h(x) = 3x^2$. ■

Example 3.   Table 12.2 gives the average weight and length for babies according to the National Center for Health Statistics.

TABLE 12.2

Age(months)	3	6	9	12	15	18	21	24
Length (in.)	24.0	26.7	28.6	30.0	31.4	32.5	33.6	34.5

Make a graph of length versus age. Explain why a power function should do a better job of fitting this plot than an exponential function. Find a power function that fits the data where age is the input and length is the output.

Solution.   Figure 12.8 gives a graph of the data from Table 12.2.

Exponential functions are always concave up. Since Figure 12.8 shows data that are concave down, a power function should be a better fit. Using power regression on a calculator, we found that our approximate regression equation was $l = 19.6a^{0.175}$, where l is length and a is age. ■

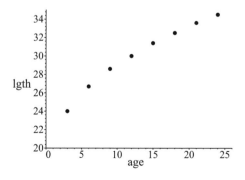

FIGURE 12.8

Application: Planets

As we noted at the beginning of this section, power functions have a number of applications ranging from traffic to area. Johannes Kepler used power functions to model the orbits of the planets. Using data similar to Table 12.3, he tried to find patterns in the movement of the planets. Table 12.3 shows the mean distance of each planet from the sun, in astronomical units (AU), and the time it takes each planet to orbit the sun, in Earth years.[5] [Note: One astronomical unit is the mean distance from the sun to Earth, approximately 1.496×10^8 km.]

TABLE **12.3**
Planet data.

Planet	Distance from the Sun (astronomical units)	Orbit time (Earth years)
Mercury	0.39	0.24
Venus	0.72	0.62
Earth	1.00	1.00
Mars	1.52	1.88
Jupiter	5.20	11.86
Saturn	9.59	29.46

As he considered the distance of each planet from the sun and the time it takes each one to orbit the sun, it was clear that the relationship was not linear since the rate of change is not constant. If we plot the data (Figure 12.9), we see that it looks like a power

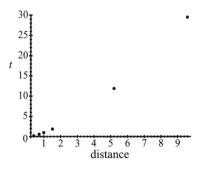

FIGURE **12.9**
A scatterplot of orbit time (in Earth years) and distance from the sun (in astronomical units) for six planets.

---

[5] *Astronomy Today*, Second Edition, Chaisson and McMillan, Prentice Hall, p. 42.

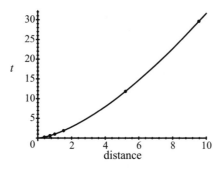

FIGURE 12.10

A plot of the six data points along with the graph of $y = x^{3/2}$.

function with an exponent greater than 1. Note that the points get close to the origin. A point at the origin would mean that some hypothetical planet would be 0 astronomical units from the sun and have an orbit time of 0 Earth years.

We can model the data with a power function by using power regression. Doing so on a calculator gives us approximately $y = x^{3/2}$. Figure 12.10 consists of a plot of the six data points along with the graph of $y = x^{3/2}$. Notice that this function is a good fit for our data.

When Kepler derived his Law, he only had the data from the first six planets in our solar system. Table 12.4 contains data for the three planets that were discovered after the time of Kepler.

TABLE 12.4

Planet data for Uranus, Neptune, and Pluto.

Planet	Date Discovered	Distance from the Sun (astronomical units)	orbit time (Earth years)
Uranus	1781	19.19	84.07
Neptune	1846	30.06	164.82
Pluto	1930	39.53	248.6

If Kepler's Law is correct, it should also work for these planets even though he did not use this data to discover the relationship between time and distance. Kepler's Law predicts that the orbit of Uranus should take about $19.19^{3/2} \approx 84.07$ years. Notice that this is Uranus's actual orbit (given in Table 12.4). So Kepler's Law appears to accurately describe the relationship between a planet's distance from the sun and the length of its orbit.

The relationship between a planet's distance from the sun and its orbit time is known as Kepler's third law of planetary motion and is usually stated as follows.

The square of the length of time it takes any planet to orbit the sun is proportional to the cube of the planet's mean distance from the sun.

Let's see if this is the same as the formula we've been using, $y = x^{3/2}$. The statement of Kepler's third law says that the two quantities are proportional. In order for one variable to be proportional to another, one variable must be equal to a constant times the other variable. In this case, Kepler's third law says that

$$t^2 = k \cdot d^3,$$

where t is time and d is distance. This can be rewritten as

$$t = k^{1/2} \cdot d^{3/2}$$

by taking the square root of both sides. Since our data contained the point $(1, 1)$, the value of $k^{1/2}$ is equal to 1. Therefore, the relationship is simply

$$t = d^{3/2},$$

the same function that we originally found.

Reading Questions for Power Functions

1. Describe some of the differences between something that can be modeled with a linear function versus something that can be modeled with a power function.

2. Below are the graphs of $y = x^{-2}$, $y = x^{-1/2}$, $y = x^{1/6}$, $y = x^{1/2}$, $y = x^2$ and $y = x^6$. Identify each, briefly justifying your answer.

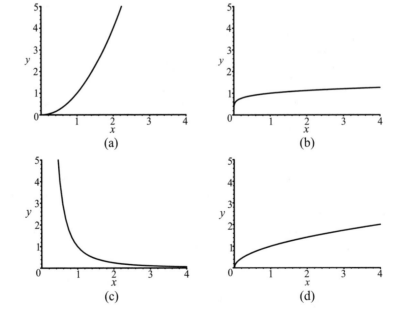

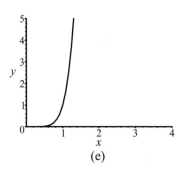

(e)

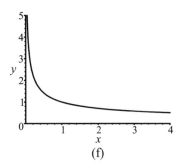

(f)

3. Let $f(x) = x^a$. Explain why it is always true that the graph passes through the point $(1, 1)$.

4. Evaluate each of the following:

   (a) $16^{1/2}$

   (b) 4^7

   (c) $125^{4/3}$

   (d) $10^{3/4}$

5. Write each of the following using radicals (i.e., write $x^{1/2}$ as $\sqrt{x}$).

   (a) $3x^{9/5}$

   (b) $x^{4/3}$

   (c) $6x^{54/5}$

   (d) $8x^{7/4}$

6. If you are given a power function whose graph is concave up in the first quadrant, what do you know about the exponent?

7. If you are given a power function whose graph is concave down in the first quadrant, what do you know about the exponent?

8. Let f be a power function of the form $f(x) = kx^p$.

   (a) Suppose $f(1) = 5$. What is the value of k?

   (b) Suppose $f(1) = 2$ and $f(2) = 8$. What is the value of p?

   (c) Suppose $f(1) = 5$ and $f(2) = 190.27$ (rounded to two decimal places). What is the value of p?

9. Below is the function whose input is the length of a pendulum and whose output is the period of the pendulum (i.e., the length of time it takes to go back and forth through one complete swing).

Length of Pendulum (in inches)	5	10	15	20	25
Period (in seconds)	0.75	1.05	1.28	1.46	1.62

    (a) Explain why this is not a linear function.

    (b) This function is a power function, $P(x) = kx^p$. Will p be greater than or less than 1? Briefly justify your answer.

10. Let f be a function whose y-intercept is 5. How do you know that this cannot be a power function?

11. Determine if each of the following is a power function. If so, give an equation for the function. If not, briefly explain why. All entries in the table are rounded to two decimal places.

(a)

x	$f(x)$
0	0
1	1
2	32
3	243
4	1024

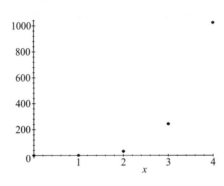

(b)

x	$f(x)$
0	0
1	4
2	16
3	36
4	64

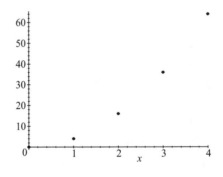

(c)

x	$f(x)$
0	0
1	1
2	1.23
3	1.39
4	1.52

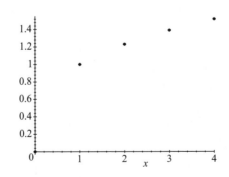

(d)

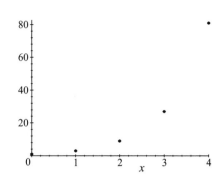

x	$f(x)$
0	1
1	3
2	9
3	27
4	81

12. Determine if each of the following is a power function by solving for y.

   (a) $x^3 = y^4$

   (b) $\frac{y}{x} = 2x^4$

   (c) $y(x - 1) = x^2 - 2x + 1$

13. Using the data from Table 12.4 in the reading, how well does Kepler's formula, $t = d^{3/2}$, work for Neptune and Pluto?

Power Functions: Activities and Class Exercises

1. **Pendulums.** The period of a pendulum (the length of time it takes to swing back and forth) depends on the length of the pendulum. For example, in a typical grandfather clock, the pendulum is about 3 feet long and takes 2 seconds to swing back and forth. A typical wall clock has a pendulum that is about 9 inches long and takes 1 second to swing back and forth. The relationship between the period and the length is a power function.

   (a) Using a pendulum, complete the second row of the following table.

Length of String (in inches)	5	10	15	20	25	30
Number of Periods (per minute)						
Period Length (in seconds)						

   (b) To find the length of a period (in seconds) you need to divide 60 by the number of periods per minute. Do this to complete the third row of the table.

   (c) Use your calculator to find a power equation that fits your data where the length of the string is your input and the period length is the output.

   (d) Theoretically, the period of a pendulum is proportional to the square root of its length. This means that the square root of the length multiplied by some constant is equal to the period. This can be written as $p = c\sqrt{l}$, where l is the length of the pendulum, c is some constant, and p is the period of the pendulum.

   i. Explain why this theoretical equation is a power function.

   ii. Does your equation fit this theoretical model? Why or why not?

(e) The constant in the theoretical model is $\frac{2\pi}{\sqrt{386}}$. So the theoretical equation is $p = \frac{2\pi}{\sqrt{386}}\sqrt{l}$. Plot your data, this equation, and the equation you found using power regression. Which equation fits your data better? Explain your answer.

2. **Beach Ball.** The height of a ball thrown up in the air can be modeled with a power function. We will look at the relationship between time and the height of a beach ball in this activity.

Directions for Using the CBR (Calculator Based Ranger)

- Plug the black cord into the the bottom of your calculator and the right-hand side of the CBR.

- If the *Ranger* program is not on your calculator, do the following steps to load it. **NOTE:** You only need to do this step if the program is not already on your calculator.

  * Press $\boxed{\text{LINK}}$ on your calculator (this is $\boxed{\text{2nd}}$ followed by $\boxed{\text{X,T,}\theta,n}$). With the arrow keys, highlight **RECEIVE**, and press $\boxed{\text{ENTER}}$. The calculator should now be "Waiting."

    Open the head of the CBR and press the $\boxed{\text{82/83}}$ button.

  * When the transfer is complete, the calculator will say DONE. If there are any problems (such as not enough memory), see your instructor for help.

  * Note: You only need to enter the program once unless you later delete it.

- Press $\boxed{\text{PRGM}}$ on your calculator. Use the arrow keys to highlight the **RANGER** program. Press $\boxed{\text{ENTER}}$. Continue to press $\boxed{\text{ENTER}}$ until you get to the screen that says **MAIN MENU** at the top.

- Choose **2:SET DEFAULTS**. Change the **Realtime** option to no, the **time(s)** option to 3, and the **UNITS:** option to feet. Do this by highlighting your selection with the arrows and pressing enter (in the case of Realtime and Units) or entering the appropriate number (in the case of time). You may need to experiment with different settings for the time. Your settings should be:

Realtime	no
time(s)	3
Display	dist
begin on	enter
smoothing	none
units	feet

- Throw the ball up in the air directly over the motion detector. **The ball must be at least 1.5 feet away from the motion detector and no more than 19 feet away from the motion detector to get a reasonable reading.** When you are ready to start, use the arrow keys to highlight **START NOW** at the top of the screen. Press $\boxed{\text{ENTER}}$ one more time and the calculator will start collecting data. It will not display the graph until it is finished taking the sample.

- Select **Plot Tools**, then **Select Domain** to "cut out" the piece of your graph that shows the path of the ball.

- Use the $\boxed{\text{TRACE}}$ key to find the y-value of the lowest and highest points on the graph showing the ball's path. Subtract to find how high the ball was thrown.

- Use the $\boxed{\text{TRACE}}$ key to find the x-value for the beginning and the end of the ball's path. Subtract to find out how long the ball stayed in the air.

- When you are finished tracing your graph, press $\boxed{\text{ENTER}}$ to get back to the **PLOT MENU**. Choose **3: REPEAT SAMPLE** to do another graph or choose **5: QUIT** if you are finished.

(a) Sketch the shape of the graph you obtained from tossing the beach ball in the air.

(b) What is the input for your function?

(c) What is the output?

(d) Use the graph to determine how high the ball was thrown. Explain how you got your answer.

(e) Use the graph to determine how long the ball was in the air. Explain how you got your answer.

3. **Land Area versus Population.** Cities often grow in terms of both land area and population. We are interested in seeing what relationship seems to best describe this growth. The following tables give the land area and population for Detroit, Seattle and Dallas.[6]

(a) For each city, find the equation of a line, exponential function, and a power function that fits the data by using LinReg, ExpReg, and PwrReg on your calculator. Record the equation and the correlation. [Note: You should have 9 equations, 3 per city.]

(b) Using the linear function for Seattle, estimate how large the city will be when the population reaches 750,000.

(c) Using the exponential function for Seattle, estimate how large the city will be when the population reaches 750,000.

------

[6] Information Obtained from the U.S. Census Bureau at: http://www.census.gov/population/documentation/twps0027/ On June 28, 1999.

(d) Using the power function for Seattle, estimate how large the city will be when the population reaches 750,000.

Detroit

Year	Land Area (sq. mi)	Population
1910	40.8	465,766
1920	77.9	993,078
1930	137.9	1,568,662
1940	137.9	1,623,452
1950	139.6	1,849,568
1960	139.6	1,670,144
1970	138.0	1,511,482
1980	135.6	1,203,339
1990	138.7	1,027,974

Seattle

Year	Land Area (sq. mi)	Population
1910	55.9	237,194
1920	58.6	315,312
1930	68.5	365,583
1940	68.5	368,302
1950	70.8	467,591
1960	88.5	557,087
1970	83.6	530,831
1980	83.6	493,846
1990	83.9	516,259

Dallas

Year	Land Area (sq. mi)	Population
1910	16.2	92,104
1920	22.8	158,976
1930	41.89	260,475
1940	40.6	294,734
1950	112.0	434,462
1960	279.9	679,684
1970	265.6	844,401
1980	333.0	904,078
1990	342.4	1,006,877

(e) Looking at the functions and the correlation, which type of function do you think best models the relationship between the size of a city and its population in general? Justify your answer.

(f) Describe several factors that influence the relationship between the land area of a city and its population.

4. **Exploring the relationship between surface area and volume.** Power functions can be used to relate surface area to volume for three-dimensional objects. One example of this is looking at the relationship between the area of a bird's wings and its weight.

(a) The formula for the volume of a cube is $V = s^3$ where s is the length of one side. The formula for the area of one side of a cube is $A = s^2$ where s is again the length of one side. Use these formulas to complete the following table.

Side Length s	Area of one face $A = s^2$	Volume $V = s^3$
1		
2		
3		
4		
5		
10		
100		
1000		
10000		

(b) Give the equation for the function that relates the area of a face of a cube to its volume. That is, determine the function where A is the input and V is the output. One way to do this is to use the data in the table from part (a). A second way is to algebraically combine the equations for $V = s^3$ and $A = s^2$. [Note: The length of the side, s, should not appear in your function.]

(c) Your answer from part (b) is a power function of the form $V = k(A^p)$. Graph this function and describe its shape.

(d) In an article entitled *Flight*, it is stated that "The fact that doubling the surface area of a cube triples its volume predicts that birds with twice the wing area will be approximately three times the weight." This statement is false. Using your function from part (b), substitute $2A$ in place of A and simplify. The volume does not triple. By what factor does the volume change?

(e) Use your calculator to make a scatterplot of the bird data given in the table on the following page.

(f) Use the power regression function on your calculator to find an equation for your data. How does this compare to the function you found in part (b)?

(g) What really seems to happen when you double the wing area of a bird? [Hint: Substitute $2x$ in place of x and simplify.]

Species	Wing Area (cm²)	Volume (grams)
Hummingbird	12.4	3.0
House Wren	48.4	11.0
Chickadee	76.0	12.5
Barn Swallow	118.5	17.0
Chimney Swift	104.0	17.3
Song Sparrow	86.5	22.0
Leach's Petrel	251.0	26.5
Purple Martin	185.5	43.0
Redwinged Blackbird	245.0	70.0
Starling	190.3	84.0

5. **Going to the Movies.** The tables on the next page list the total revenue for three different movies. The week indicates the beginning of the week (i.e. Week 1 represents the opening weekend) and the total revenue is in millions of dollars.[7] For each of these, do the following:

- Make a scatterplot of the data.
- Use power regression on your calculator to determine the equation of the power function that best fits the data. (See calculator instructions for TI-83 in the appendix.)
- Sketch the graph of your function on the scatterplot.

(a) For each of the three movies, write the equation of the power function, $y = kx^p$. The input should be *weeks* and the output should be *total revenue*.

(b) In terms of the movie data, what is the meaning of k?

(c) Looking at the graph, when should a company decide to pull a movie from the theaters?

(d) Suppose the power function modeling the revenue of movie A was $f(x) = 45x^{0.56}$ while the power function modeling the revenue of movie B was $g(x) = 66x^{0.4}$.

   i. Which movie had the better opening weekend?

   ii. If the movies stay in the theater for 10 weeks, which one makes more money?

   iii. If the movies stay in the theater for 20 weeks, which one makes more money?

(e) What does it mean if the exponent of the power function that models the revenue from a movie is greater than 1?

----

[7] Internet Movie Database, 7 March 2000, <http://us.imdb.com>.

Titanic	
Opened Dec 21, 1997	
Weeks	Total Revenue
1	$28.638
2	$88.425
3	$157.467
4	$197.881
5	$242.748
6	$274.599
7	$308.100
8	$337.355
9	$376.270
10	$402.561
11	$426.983
12	$449.157
13	$471.446
14	$494.514
15	$515.262
16	$530.406
17	$542.853
18	$554.067
19	$560.615
20	$565.736
21	$569.820
22	$572.713
23	$577.060
24	$579.419
not in theater	
not in theater	
not in theater	
not in theater	
not in theater	
not in theater	
not in theater	
not in theater	

Star Wars I	
Opened May 23, 1999	
Weeks	Total Revenue
1	$105.661
2	$207.099
3	$255.758
4	$296.964
5	$328.072
6	$351.669
7	$373.166
8	$385.184
9	$395.201
10	$402.770
11	$408.597
12	$412.775
13	$415.645
14	$417.800
15	$419.423
16	$421.381
17	$422.252
18	$423.299
19	$424.399
20	$425.360
21	$426.141
22	$426.818
23	$427.285
24	$427.607
25	$429.002
26	not available
27	$429.984
28	$430.443
29	$430.727
30	$430.898
31	$430.984
32	$431.065

The Matrix	
Opened April 4, 1999	
Weeks	Total Revenue
1	$37.352
2	$73.310
3	$98.946
4	$117.082
5	$129.715
6	$138.505
7	$145,138
8	$149.507
9	$154.764
10	$158.260
11	$161.367
12	$163.869
13	$165.688
14	$166.784
15	$167.505
16	$167.910
17	$168.220
18	$168.932
19	$169.731
20	$170.416
21	$171.051
22	$171.229
23	$171.321
24	$171.383
not in theater	
not in theater	
not in theater	
not in theater	
not in theater	
not in theater	
not in theater	
not in theater	

6. **Baby Weights.** The following table gives the average weight for babies according to the National Center for Health Statistics.

Age (months)	3	6	9	12	15	18	21	24
Weight (lb)	12.3	16.2	19.1	21.1	22.8	24.0	25.0	25.8

(a) Make a graph of weight versus age. (Your graph should have age on the horizontal axis.)

(b) By looking at your graph, explain why a power function should do a better job of fitting these data than an exponential function.

(c) Using power regression, find a power function that fits the data where age is the input and weight is the output.

(d) Plot your power function along with the data. How well does your function fit the data?

(e) The average weight of a one month old baby is 9.7 pounds. Using your power equation, what is the average weight of a one month old baby?

(f) The average weight of a newborn baby is 7.9 pounds. Using your power equation, what is the average weight of a newborn baby?

(g) The following table, taken from the National Center for Health Statistics, gives the average weight of children ages 3 through 10.

Age (years)	3	4	5	6	7	8	9	10
Weight (lb)	32	36	40	46	51	57	63	70

Plot these data along with your power function and the original data from part (d). In doing so, remember to convert the years to months. For example, the first new point should be 36 months with a weight of 32 pounds.

(h) How well does your function fit the new data points from part (g)? How does this compare with the fit of the original data points?

13

Probability

Buying a lottery ticket, trading in the stock market, or just deciding whether to take an umbrella on a cloudy day are dependent on your estimate of probable outcomes. The branch of mathematics that studies the outcomes of random events is called probability. We will start with some techniques that can be used to systematically find all possible outcomes in a given situation. We will then discuss the idea of randomness as it relates to those outcomes. Finally, we will see how to set up simple probability models and how to describe random events using a probability distribution.

Counting and Listing All Outcomes

Odometer Method

We need to be able to list and count all possible outcomes before we can talk about how to compute probabilities. For example, suppose a student is going to a fast food restaurant for lunch. The student likes hamburgers, chicken sandwiches, and fish sandwiches. He likes french fries and onion rings. He likes coke, cherry coke, sprite and tea. We want to know how many different lunch combinations are acceptable to this student. A lunch combination has one sandwich, either french fries or onion rings, and one drink. We will first list all the possible food combinations and then we will count all the possible combinations. We will list all the possible food combinations by using a technique called the "Odometer Method." Think of the odometer on a car. It displays a series of numbers which show how far the car has traveled. For illustration, we will consider an odometer with three positions where each position is a number between 0 and 9 (10 possible numbers).

Left-most position (hundreds)	Middle position (tens)	Right-most position (units)
0, 1, 2, 3, 4, 5, 6, 7, 8, 9	0, 1, 2, 3, 4, 5, 6, 7, 8, 9	0, 1, 2, 3, 4, 5, 6, 7, 8, 9
choose one	*choose one*	*choose one*

At the beginning, the odometer reads 000. When the car has traveled 1 mile, the reading is 001. When the car has traveled 9 miles, the reading is 009. When the car reaches 10 miles, the odometer reads 010. Think about what actually happened. The right-most position on the odometer has gone from 0 to 9 then back to 0. It cycled through all 10 of its possible numbers. Meanwhile, the middle position on the odometer has cycled from 0 to 1. It will stay at 1 until the right-most number fully cycles again, then it will move to 2. If we were to try and list the odometer reading after each mile of travel we could generate a list that would look like this:

000	001	002	003	004	005	006	007	008	009
010	011	012	013	014	015	016	017	018	019
020	021	022	...						

$\vdots$

570 571 572 573 ...

$\vdots$

990 991 992 993 994 995   ...   999 (a big long list)

Note that if we look at the three numbers showing on the odometer, no reading is repeated in our list. For example, the reading 572 appears only once in our list. How does this help us list all the food combinations? Just as our odometer has three slots or positions that are to be filled with numbers, we can think of food combinations as filling three slots with the student's choices. These will be the sandwich slot, the side order slot and the drink slot. For example, one odometer reading would be as follows.

hamburger   french fries   coke
(sandwich)   (side order)   (drink)

We could say that the reading (hamburger, french fries, coke) means the student ordered this combination for his lunch.

We list the possibilities as follows:

Sandwich	**Side Item**	**Drink**
hamburger, chicken, fish	french fries, onion rings	coke, cherry coke, sprite, tea
choose one	*choose one*	*choose one*

To list all the possible ways the student can order lunch, we will cycle through the possibilities just like the odometer. Table 13.1 shows that our list contains 24 possible food combinations.

We first chose the left-most choice in each category: hamburger, french fries, and coke. We then kept the sandwich choice (hamburger) and the side item choice (french fries) fixed while we cycled through each of the drink choices. When this was completed, we changed the side item to onion rings and again cycled through each of the drinks. It wasn't until we cycled through all the side order choices that we then changed the sandwich order to chicken.

TABLE **13.1**

Twenty-four possible lunch combinations

1. hamburger, french fries, coke	2. hamburger, french fries, cherry coke
3. hamburger, french fries, sprite	4. hamburger, french fries, tea
5. hamburger, onion rings, coke	6. hamburger, onion rings, cherry coke
7. hamburger, onion rings, sprite	8. hamburger, onion rings, tea
9. chicken, french fries, coke	10. chicken, french fries, cherry coke
11. chicken, french fries, sprite	12. chicken, french fries, tea
13. chicken, onion rings, coke	14. chicken, onion rings, cherry coke
15. chicken, onion rings, sprite	16. chicken, onion rings, tea
17. fish, french fries, coke	18. fish, french fries, cherry coke
19. fish, french fries, sprite	20. fish, french fries, tea
21. fish, onion rings, coke	22. fish, onion rings, cherry coke
23. fish, onion rings, sprite	24. fish, onion rings, tea

Counting all outcomes

Suppose we wanted to know how many lunch combinations are possible without actually listing them. We know that the student has 3 choices for sandwiches and 2 choices for side item. If we consider just sandwich and side item combinations, we see that the hamburger can be paired with either french fries or onion rings, the chicken sandwich with either french fries or onion rings, or the fish with either french fries or onion rings. There are thus 6 possible sandwich/side item combinations (3 sandwiches, 2 side orders giving $3 \times 2 = 6$ combinations). Each of these 6 combinations can be paired with one of the 4 drinks giving 24 possible outcomes ($6 \times 4 = 24$). This gives us a **general counting method**. When faced with a sequence of choices (such as choosing a sandwich, then a side item and finally a drink), the number of different combinations is the product of the number of ways each choice can be made ($3 \times 2 \times 4 = 24$).

Equally Likely Outcomes

If you drop a quarter into a slot machine at a casino, the outcome is uncertain. However, if hundreds of people put several quarters in a slot machine, the outcome becomes quite predictable—the casino will make money. Using the laws of probability, even how much profit the casino will make can be accurately predicted. Predicting these types of results involves looking at the event in terms of equally likely outcomes.

Finding probabilities of equally likely events is connected with finding the sample space. The **sample space** is the set of all equally possible outcomes. Suppose, for example, that the lunch choices for our student are equally likely. This means that the choices are made randomly (by a method such as flipping a coin or rolling a die so that any one choice is just as likely as any other). The sample space for the lunch example is the list of the 24 possible combinations in Table 13.1.

An **event** is a subset of outcomes of the sample space. Thus, an event is a collection of outcomes selected from the sample space. For example, "eating a hamburger" is an

event for our lunch example. The **probability of an event** is a measure of how likely
the event is to occur. When each item in the sample space has an equally likely chance
of occurring, the probability of an event is defined as

$$\frac{\text{the number of outcomes of an event}}{\text{the total number of outcomes in the sample space}}.$$

So the probability of an event has become a counting problem. If our event is "eating a
hamburger," we need to count the outcomes containing a hamburger and divide that by 24
(the total number of outcomes for our lunch combinations). Referring back to Table 13.1,
we see that there are 8 combinations containing a hamburger. So the probability of eating
a hamburger is 8 divided by 24 or 1/3.

Many probability problems deal with dice or cards. In fact, the foundations of
probability were the result of determining how to fairly divide the winnings in a game of
chance. Suppose you were determining probabilities for rolling a pair of dice. To answer
this question, we need to start by finding the number of equally likely outcomes for the
sum. You might guess that, since you could have a sum of 2, 3, 4, 5, 6, 7, 8, 9, 10, 11, or
12 showing on the two dice, there are eleven outcomes. However, these outcomes are not
equally likely. There is only one way to get a sum of two (rolling two ones) while there
are many ways to get a sum of seven (rolling a three and a four, two and a five, etc).
To predict what will happen if the process is repeated many times, the process must be
broken down into equally likely outcomes. For rolling two dice, one gray and one white
for example, there are 36 equally likely outcomes. (See Figure 13.1.)

From Figure 13.1, you can see that since there is one way to get a sum of 2 and 36
equally likely outcomes, the probability of rolling a 2 is $\frac{1}{36}$. You can also see that there
are six ways to roll a sum of 7 (these show up on the diagonal running from the bottom
left to the upper right). Therefore, the probability of rolling a 7 is $\frac{6}{36} = \frac{1}{6}$. (Notice that
getting a total of 7 when the gray die comes up 3 and the white die comes up 4 is a
different outcome from the gray die coming up 4 and the white die coming up 3.)

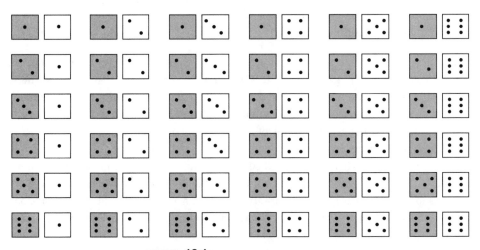

FIGURE 13.1
The sample space for rolling two dice.

An event can be more complicated than rolling a given sum. For example, a possible event is rolling a sum that is less than 4. To find the probability of this event, we use Figure 13.1 to count the number of outcomes that give a sum which is less than 4. Since there are three of these (rolling two 1s, rolling a 1 and a 2, or rolling a 2 and a 1), the probability of the event is $\frac{3}{36} = \frac{1}{12}$.

Another term frequently used in probability is complement. The **complement** of an event is the set of all outcomes in the sample space that are not in the event. For example, the complement of rolling a sum that is less than 4 is rolling a sum that is 4 or greater. *Sample space*, *event* and *complement* refer to the outcomes of a random phenomenon rather than the associated probabilities. One way to compute the probability is to first determine the outcomes that make up the event.

Because of its definition, the probability of an event will always be greater than or equal to 0 and less than or equal to 1. It must be nonnegative because counting outcomes is never negative. It must be less than or equal to 1 because the number of outcomes in the event can never exceed the number of outcomes in the sample space. A probability of 0 means that the event will never happen and a probability of 1 means that it will always happen. Since an event and its complement constitute the entire sample space, the sum of the probabilities of an event and its complement is 1. This is expressed in symbols as $P(E) + P(E') = 1$ where E' refers to the complement of event E and the P represents "probability of." This property can be quite useful. It is sometimes easier to compute the probability of the complement of an event rather than the probability of the actual event itself. For example, suppose you rolled two dice and were asked to find the probability of rolling a sum other than 2. You could count all of the ways to not roll a 2, or you could determine the probability of its complement, which is rolling a sum of 2. This probability is $\frac{1}{36}$. Since the event and its complement sum to 1, we subtract $\frac{1}{36}$ from 1 to find that the probability of not rolling a 2 is $1 - \frac{1}{36} = \frac{35}{36}$.

To summarize these two properties, we have:

- The probability of any event, $P(E)$, is a number between 0 and 1, i.e., $0 \le P(E) \le 1$.

- For any event, E, and its complement, E', we have $P(E) + P(E') = 1$. This means that $P(E') = 1 - P(E)$.

Multi-stage Experiments

Determining the probability of an event by first writing down all the outcomes of the sample space and then counting the number of outcomes in the event works well when the random phenomenon is fairly simple and the number of outcomes is fairly small. When events become more complex, we use properties which make this computation easier. One such property is:

> *The probability of two or more independent events is the product of the associated probabilities.*

For example, suppose we roll a die twice and want to calculate the probability of rolling a 2 on the first roll and an even number on the second roll. We could look at the sample space in Figure 13.1 and count the number of outcomes for which this is true. If we think of the number on the left in each pair as the outcome for the first roll and the number on the right as the outcome for the second roll, we see that there are three outcomes in which the first roll is a 2 and the second is even. These are $(2, 2)$, $(2, 4)$, and $(2, 6)$. Therefore, the probability of this event happening is $\frac{3}{36} = \frac{1}{12}$. This result could also be computed by multiplying the probability of the first event by the probability of the second event. The probability of rolling a 2 on the first roll is $\frac{1}{6}$. The probability of rolling an even number on the second roll is $\frac{3}{6}$ since there are three even numbers on a die. We multiply the probabilities of these two events to determine the probability of both events happening. Therefore, the probability of rolling a 2 on the first roll and an even number on the second is $\frac{1}{6} \cdot \frac{3}{6} = \frac{3}{36} = \frac{1}{12}$. Observe that $\frac{1}{6}$ of the outcomes in Figure 13.1 have a 2 on the first die and $\frac{3}{6} = \frac{1}{2}$ of those outcomes have an even number on the second die. Therefore, $\frac{1}{6}$ of $\frac{1}{2}$ is $\frac{1}{6} \cdot \frac{1}{2} = \frac{1}{12}$.

You can only multiply two probabilities to find the probability of both things happening if the two events are independent. Two events are **independent** if the occurrence of one does not change the probability of the other occurring. Rolling two dice produces independent events because the outcome of the first die is independent of the outcome of the second. However, not all events are independent. For example, the probability of you having type A blood is *dependent* on whether or not either of your parents has type A blood. The probability that tomorrow is sunny depends on today's weather.

Two French mathematicians, Blaise Pascal (1623–1662) and Pierre Fermat (1601–1665), are credited for developing the first theories of probability. A problem was proposed to Pascal by Antoine Gombaud, a professional gambler, who asked:

> If two players were playing a game, such as dice, and the game was cut short before it was concluded, how should the winnings be divided?

Pascal and Fermat began to develop the early laws of probability as they corresponded about how to answer this question.

One of the questions Pascal and Fermat considered was: "What is the probability of throwing at least one 6 in four rolls of a die?" To solve this problem we will use the properties of probability we have developed. Looking at the problem directly is too complicated since rolling at least one 6 means we have to consider the events for rolling exactly one 6, exactly two 6s, exactly three 6s, or exactly four 6s. However, the complement of rolling at least one 6 in four rolls of the die is rolling no 6s. This is much easier to solve. Rolling the die four times can be thought of as four independent events. This means we can multiply the associated probabilities. Since the probability of not rolling a 6 on a single die is $\frac{5}{6}$, the probability of rolling a die four times without rolling a 6 is

$$\frac{5}{6} \cdot \frac{5}{6} \cdot \frac{5}{6} \cdot \frac{5}{6} = \left(\frac{5}{6}\right)^4 \approx 0.48$$

Therefore the probability of rolling at least one 6 is

$$1 - \left(\frac{5}{6}\right)^4 \approx 0.52.$$

Since $0.52 > \frac{1}{2}$, we know that when rolling a die four times (or rolling four dice at the same time), it is more likely than not that at least one of the rolls will be a 6.

Probability Distributions

A **probability distribution** is the collection of all the outcomes of a random phenomenon together with their associated probabilities. For example, a probability distribution for rolling two dice and recording the sum is shown in Table 13.2. The probability distribution shown in Table 13.3 lists the different colors of M&M's and the corresponding probability that you would randomly choose that color from a bag of M&M's. Notice that the sum of the probabilities in a probability distribution is always 1.

TABLE **13.2**
The probability distribution for rolling two dice and recording the sum.

Outcome: Sum	2	3	4	5	6	7	8	9	10	11	12
Probability	$\frac{1}{36}$	$\frac{2}{36}$	$\frac{3}{36}$	$\frac{4}{36}$	$\frac{5}{36}$	$\frac{6}{36}$	$\frac{5}{36}$	$\frac{4}{36}$	$\frac{3}{36}$	$\frac{2}{36}$	$\frac{1}{36}$

TABLE **13.3**
The probability distribution for colors of milk chocolate M&M's .

Outcome	Brown	Yellow	Red	Orange	Green	Blue
Probability	0.3	0.2	0.2	0.1	0.1	0.1

One difference between the probability distributions in Table 13.2 and Table 13.3 is that the mean of the distribution can be found for rolling two dice (Table 13.2), but not for colors of M&M's (Table 13.3). This is because the outcomes in Table 13.2 are represented by numbers. The outcomes in Table 13.3 are not represented by numbers, so it does not make sense to find the mean of M&M colors. To find the **mean of a probability distribution**, you multiply the outcomes by the respective probability and add the products. For example, the mean of the probability distribution in Table 13.2 is

$$2 \cdot \frac{1}{36} + 3 \cdot \frac{2}{36} + 4 \cdot \frac{3}{36} + 5 \cdot \frac{4}{36} + 6 \cdot \frac{5}{36} + 7 \cdot \frac{6}{36} + 8 \cdot \frac{5}{36}$$

$$+ 9 \cdot \frac{4}{36} + 10 \cdot \frac{3}{36} + 11 \cdot \frac{2}{36} + 12 \cdot \frac{1}{36} = 7.$$

Suppose three coins are flipped and you record the number of tails. The possibilities are 0 tails, 1 tail, 2 tails, or 3 tails. There are eight different outcomes when flipping three coins: HHH, HHT, HTH, THH, HTT, THT, TTH, TTT. (Think of the coins as a penny, a nickel, and a dime, so heads on the penny and the nickel and tails on the dime, HHT, is different from heads on the penny and dime and tails on the nickel, HTH.) This

TABLE **13.4**

The probability distribution for the number of tails observed when tossing three coins.

Number of Tails	0	1	2	3
Probability	$\frac{1}{8}$	$\frac{3}{8}$	$\frac{3}{8}$	$\frac{1}{8}$

is a sample space of size 8. Looking at the sample space, you can see that there is 1 way to get 0 tails, 3 ways to get 1 tail, 3 ways to get 2 tails and 1 way to get 3 tails. The probability distribution for flipping 3 coins and recording the number of tails is shown in Table 13.4.

We want to find the mean number of tails recorded when tossing three coins. One method is to look at the eight different outcomes, add up the number of tails for each outcome, and divide by 8:

$$\text{mean number of tails} = (0 + 1 + 1 + 1 + 2 + 2 + 2 + 3) \div 8 = \frac{12}{8} = 1.5.$$

We could also find the sum of the products of each outcome and its associated probability to get

$$\text{mean} = 0\left(\frac{1}{8}\right) + 1\left(\frac{3}{8}\right) + 2\left(\frac{3}{8}\right) + 3\left(\frac{1}{8}\right) = \frac{0}{8} + \frac{3}{8} + \frac{6}{8} + \frac{3}{8} = \frac{12}{8} = 1.5.$$

These methods accomplish the same thing. The second method simply divides each term by 8 before adding rather than afterwards.

The state of Michigan has a twice-weekly Lotto game in which the player picks six numbers from the set of integers 1–49. The state then picks six. You win a prize if you match all six, five out of six, or four out of six. The probability distribution where the outcomes are the number of matches with their associated probabilities is shown in Table 13.5.

Notice that the probability of matching all six numbers is $\frac{1}{13,983,816}$. The denominator of this fraction is so large that it is hard to put it in perspective. Imagine yourself in a long line waiting to buy a single lotto ticket. One person in that line will be the winner. The probability of matching all six numbers, $\frac{1}{13,983,816}$, means you are standing in a line that is 13,983,816 people long. Such a line would extend from New York City to Los Angeles and back! Only one person will be picked from that line of almost 14 million people to win the lotto jackpot. That is your probability of winning the lotto.

TABLE **13.5**

The probability distribution for the number of matches in the Michigan Lotto.

No. of matches	6	5	4	3	2	1	0
Prob.	$\frac{1}{13,983,816}$	$\frac{258}{13,983,816}$	$\frac{13,545}{13,983,816}$	$\frac{246,820}{13,983,816}$	$\frac{1,851,150}{13,983,816}$	$\frac{5,775,588}{13,983,816}$	$\frac{6,096,454}{13,983,816}$

To find out how many numbers you would expect to match on average if you bought a ticket, we find the mean of the distribution given in Table 13.5:

$$6\left(\frac{1}{13,983,816}\right) + 5\left(\frac{258}{13,983,816}\right) + 4\left(\frac{13,545}{13,983,816}\right) + 3\left(\frac{246,820}{13,983,816}\right)$$
$$+ 2\left(\frac{1,851,150}{13,983,816}\right) + 1\left(\frac{5,775,588}{13,983,816}\right) + 0\left(\frac{6,096,454}{13,983,816}\right) \approx 0.73.$$

This means that, on average, you will match 0.73 of the winning numbers. In other words, you can expect to match less than one number, on the average. Therefore, it should not surprise you if you buy a Lotto ticket and do not match a single number.

Instead of calculating the number of matching numbers, on average, we could calculate the money won, on average, from playing lotto. The mean of a probability distribution involving money is called the **expected value**. The state gives $2 million dollars to a winner matching all six numbers. If you match five out of six you win $2500, if you match four out of six you win $100, and for anything else, you win nothing. The probability distribution for the amount you could win listed as the outcomes is shown in Table 13.6.

TABLE 13.6
The probability distribution for the dollar value of the winnings in the Michigan Lotto.

Winnings	$2,000,000	$2500	$100	$0	$0	$0	$0
Prob.	$\frac{1}{13,983,816}$	$\frac{258}{13,983,816}$	$\frac{13,545}{13,983,816}$	$\frac{246,820}{13,983,816}$	$\frac{1,851,150}{13,983,816}$	$\frac{5,775,588}{13,983,816}$	$\frac{6,096,454}{13,983,816}$

To calculate how much you could expect to win if you bought a ticket, we find the mean of this probability distribution.

$$\$2,000,000\left(\frac{1}{13,983,816}\right) + \$2500\left(\frac{258}{13,983,816}\right) + \$100\left(\frac{13,545}{13,983,816}\right) + \$0\left(\frac{246,820}{13,983,816}\right)$$
$$+ \$0\left(\frac{1,851,150}{13,983,816}\right) + \$0\left(\frac{5,775,588}{13,983,816}\right) + \$0\left(\frac{6,096,454}{13,983,816}\right) \approx \$0.28$$

The expected value for playing the Michigan Lotto is 28 cents. This means that you could expect to win, on average, just 28 cents per ticket. Because it costs $1 for each ticket, you lose 72 cents per ticket on average. So, if you bought 100 tickets, costing you $100 total, you should expect to win only about $28. This is not a particularly good deal for the overwhelming majority of lotto players, but it certainly helps fill the state's coffers.

Application: Streaks

In 1941, Joe DiMaggio got a hit in 56 consecutive baseball games while playing for the New York Yankees. Many call this not only the greatest streak in baseball, but the greatest streak in all sports. Just how improbable is it for a hitter similar to DiMaggio to get a hit in 56 consecutive games?

Example 1.   In 1941, Joe DiMaggio had a batting average of 0.357. We will assume that for each at bat, the probability that he would get a hit was 0.357. He had slightly fewer than four at bats per game. We will assume he had exactly four at bats each game. What is the probability that Joe DiMaggio would get at least one hit in a game? What is the probability that Joe DiMaggio would get at least one hit in the next 56 consecutive games?

Solution.   To find the probability that DiMaggio would get at least one hit in a game, we need to find the probability that he would get exactly one, exactly two, exactly three, or exactly four hits. Since the only other thing that can happen (or the complement of this event) is getting no hits, we will calculate that probability and subtract it from 1. Therefore, the probability that DiMaggio would not get a hit in a single at bat is

$$1 - 0.357 = 0.643.$$

The probability that DiMaggio would get no hits in four at bats, is

$$(0.643)(0.643)(0.643)(0.643) = 0.643^4 \approx 0.171.$$

Hence, the probability that DiMaggio would get at least one hit in a game is

$$1 - 0.171 = 0.829.$$

   Now we can use our answer from the first part to find the probability that DiMaggio would get at least one hit in 56 games in a row. To do this we just need to multiply 0.829 by itself 56 times to get

$$0.829^{56} \approx 0.0000275.$$

Therefore, the probability that DiMaggio would get at least one hit in 56 games in a row is 0.0000275, a very unlikely event. ■

   In the previous example, we see that the probability of getting at least one hit in 56 games in a row is 0.0000275. You can think of this as the probability of getting at least one hit in 56 games in a row if there were only 56 games in the season. There were many more than that however. Therefore the probability that DiMaggio would get a hit in 56 games in a row sometime during the 1941 season is higher. Also, the probability that DiMaggio would get a hit in 56 games in a row sometime during his career is even higher yet. Even so, no one has even come close to DiMaggio's streak. The next closest is 44 in a row. This was accomplished by Willie Keeler in 1897 and Pete Rose in 1978.

Reading Questions for Probability

1. Suppose a new car model comes with the following option: 3 choices of body style (2-door, 4-door or station wagon), 4 colors (black, red, green or white), and 2 choices of engine size (4 or 6 cylinder). List all the different ways the model could be ordered starting with (2,B,4) for a 2-door black 4-cylinder model.

2. How many different lunch choices would there be if you choose from 4 sandwiches, 2 side orders, and 5 drinks? [Note: You do not need to list all of the possible combinations.]

3. Why can a probability never be greater than 1?

4. If the probability it will rain tomorrow is 0.6, what is the probability of the complement of that event?

5. Suppose you are tossing 3 coins and want at least 2 heads.

   (a) List all of the possible outcomes for the sample space.

   (b) List all of the possible outcomes for the event.

   (c) What is the probability of tossing 3 coins and having at least 2 heads?

6. When rolling two dice, what is the probability that the sum is greater than or equal to 7, but less than 10?

7. Suppose you have a jar with 5 white marbles and 8 black marbles.

   (a) What is the probability of drawing a black marble?

   (b) What is the probability of drawing a white marble?

   (c) Find the sum of your answers from parts (a) and (b). Why is this true?

   (d) Suppose you draw one marble, replace it, and draw a second marble.

      i. What is the probability that both marbles are black?

      ii. What is the probability that both marbles are white?

      iii. What is the probability that the first marble is black and the second one is white?

8. When rolling two dice, what is the probability that you get a 3 on the first die and a number greater than 3 on the second?

9. What is a probability distribution?

10. Give an example, not mentioned in the reading, of a probability distribution where it is possible to find the mean.

11. Give an example, not mentioned in the reading, of a probability distribution where it is not possible to find the mean.

12. List the sample space for the experiment in which four coins are flipped. (Hint: There are 16 different outcomes.)

13. When flipping four coins, determine the following probabilities.

   (a) Getting four heads.

   (b) Getting at least three heads.

   (c) Getting a head on the second coin.

14. Find the mean of the following probability distribution where the outcome is the number of people in a family. Describe what your answer means.

Persons in Family	2	3	4	5	6	7
Probability	0.41	0.24	0.21	0.09	0.03	0.02

15. Aunt Hazel's Raisin Cookies were analyzed for their number of raisins. The following probabilities were found:

Number of Raisins, x	3	4	5	6	7	8
Probability of x Raisins	0.10	0.15	0.15	0.25	0.20	0.15

Find the mean number of raisins in a cookie.

16. Calvin is taking a test using the "Calvin Method" which involves flipping a coin and putting "true" for heads and "false" for tails. Suppose the test contains ten questions. What is the probability that he gets all ten questions right?

17. A family has three children.

    (a) What is the probability that the two oldest are girls and the youngest is a boy?

    (b) What is the probability that there are two girls and one boy (in any order)?

18. Suppose that, when throwing darts, you make a bull's eye 40% of the time. If you throw two darts, what is the probability that neither of them is a bull's eye?

19. You and a friend decide to play "gambling dice." It costs 1 dollar to play. If you roll a 5 or 6, you get 2 dollars (a 1 dollar profit). If not, you lose your dollar. Compute the expected value for this game.

20. In 1978, Pete Rose got a hit in 44 consecutive baseball games. This hitting streak is second only to that of Joe DiMaggio's 56 game hitting streak in 1941. In 1978, Rose had a batting average of 0.302. We will assume that for each at bat, the probability that he would get a hit was 0.302. He had slightly more than four at bats per game in 1978. We will assume he had exactly four at bats each game.

    (a) What is the probability that Pete Rose would get at least one hit in a game?

    (b) What is the probability that Pete Rose would get at least one hit in the next 44 consecutive games?

Probability: Activities and Class Exercises

1. **Mini Lotto.** In Mini Lotto, you try to match the two numbers that are drawn. It costs $1 to play and you receive $5 if you win. You can only choose numbers from 1 to 6. Numbers are not repeated so be sure to choose two *different* numbers. Also, order does not matter (i.e., if the numbers drawn are $(1, 2)$, you win regardless of whether you choose 1, then 2; or 2, then 1).

    (a) Using one die, play the game five times, keeping track of how much money you won or lost in the following table. Assume you start with $5. Be sure to subtract a dollar every time you play.

Total Money	Numbers Picked	Numbers Drawn	Amount won/lost
$5			

(b) List all possible combinations of numbers that can be drawn playing Mini Lotto. [Hint: There are 15 of these.]

(c) What is the probability that you will win Mini Lotto? What is the probability that you will match one number in Mini Lotto?

(d) Compute the expected value for Mini Lotto.

2. **Garage Door.** Many people use remote control garage door openers. Since there are limitations on the number of possible codes, not everyone will have a different code. Our goal in this problem is to determine how likely it is that your garage door can be opened by someone else's remote unit. A remote control unit opens a garage door by emitting a radio frequency. A range of frequencies is programmed into these remote control units. You can then select your frequency by inserting a code both into your remote unit and the receiver in the garage. A common method is to use a series of switches that can be turned on or off. The arrangement of these switches determines the code. If a remote unit had one switch, there would be two codes: {off, on}. If there were two switches, there would be $2^2 = 4$ codes: {off-off, off-on, on-off, on-on}. If there were three switches, there would be $2^3 = 8$ codes: {off-off-off, on-off-off, off-on-off, off-off-on, off-on-on, on-off-on, on-on-off, on-on-on}.

(a) Suppose a garage door opener has a remote with ten switches. Each of these switches can be turned on or off. The arrangement of these switches determines the code for that particular garage door opener.

   i. How many different arrangements are there for these ten switches?

   ii. If your next door neighbor has the same type of garage door opener[1] and both of you had your codes randomly set at the factory, what is the probability that they are set on the same frequency?

(b) While the probability that two houses share the same garage door code is quite small, the problem we are considering is whether a given code matches *any* of the neighbors' codes. We are interested in finding the probability that a given opener will open *at least* one of the other doors in a certain subdivision. This is equivalent to finding the probability of matching the code of exactly one other remote, or exactly two others, or exactly three, or ... or all the others. In this

---

[1] This is likely since neighboring houses are often built by the same builder during the same year.

case, it is easier to find the probability of the complement, which is that the given code matches exactly zero of the others.

Let's say you live in a subdivision where there are seventeen homes. Assume each home was built in the past two years, most by the same contractor. Assume they all have the same model of garage door opener, that each home has exactly one garage door opener, and that the codes were randomly set at the factory. Assume you live in lot #1. (See the following figure.)

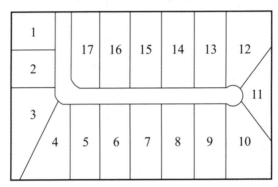

i. What is the probability that your code does *not* match the code in lot #2? Remember there are ten switches in each remote.

ii. What is the probability that your code does *not* match any of the 16 other homes in the subdivision?

iii. What is the probability that your code *does* match at least one of the other 16 homes in the subdivision?

(c) Problems in the neighborhood could arise not only if *your* garage door code matches another, but if any code matches any other. So we need to go back to the subdivision and find the probability that there is *at least* one match between any of the 17 codes. Because this question is concerned with *at least* one match, we will again use the complement. Let's repeat some of our earlier work, concentrating on the pattern that develops. Don't forget, you live in lot #1.

i. What is the probability that the code in lot #2 does not match yours?

ii. Assuming the code in lot #2 is different from yours, what is the probability that the code in lot #3 does not match either the code in lot #2 or yours? [Note: There are now two fewer unused codes for the person in lot #3 to have.]

iii. Assuming that your code, the code in lot #2, and the code in lot #3 are all different, what is the probability that the neighbor in lot #4 has a code that does not match either the code in lot #3, the code in lot #2, or your code?

iv. Multiply your answers from (c)i, (c)ii, and (c)iii to find the probability that there is no match among the four of you.

v. You should now see a pattern develop as you start including other neighbors in this scenario. Using this pattern, determine the probability that no one among the 17 houses in your subdivision has a code that matches anyone else's.

  vi. Looking at the complement, what is the probability that at least one member of your subdivision has a code that matches another?

(d) Do you think that if a garage door opener has ten switches, there are enough different codes so that any remote in this subdivision is unlikely to open someone else's garage door? If not, how might you change the code to make it more unlikely that this problem doesn't occur?

3. **Sports.** In professional sports, there are different types of playoff systems. In football, the winner of one game advances to the next round. However in baseball, the playoff to get into the World Series is a best-of-seven game series. In this problem we will look at the mathematics of these types of series.

(a) Suppose you and a friend play ping-pong frequently. You know your friend is a little better than you at this game since he or she wins more often. You decide to play a tournament and the loser has to buy the pizza for the night. Knowing that you are not quite as good as your friend, would you want to play one game—winner take all, or play a best of three-game series? Justify your answer.

(b) Suppose that instead of you and a friend playing ping-pong, there are two schools that play each other in basketball. We will call these two schools Harvard and Columbia. Also suppose Harvard, being the better team, has a probability of $\frac{2}{3}$ of winning any particular game. When these two teams play, we would usually expect the better team to win. Through simulation, we will see if the better team is more likely to win in a one game series, a best-of-three game series, a best-of-five game series, or a best-of-seven game series. Your calculator can be used to simulate this event by having it choose random integers between 1 and 3. Each random integer will represent the outcome of a basketball game. If the random number is a 1, then mark down Columbia as the winner. If the random number is a 2 or a 3, then mark down Harvard as the winner.

  i. Have your calculator "play" 25 individual games and record the number of Harvard wins and the number of Columbia wins. What percent of the games did Harvard win? Remember, if the random number is a 1, then mark down Columbia as the winner. If the random number is a 2 or a 3, then mark down Harvard as the winner.

  ii. Have your calculator "play" 25 three-game-series and record the number of Harvard series wins and then number of Columbia wins. What percent of the three-game-series did Harvard win?

  iii. Have your calculator "play" 25 five-game-series and record the number of Harvard series wins and the number of Columbia series wins. What percent of the five-game-series did Harvard win?

  iv. Have your calculator "play" 25 seven-game-series and record the number of Harvard series wins and the number of Columbia series wins. What percent of the seven-game-series did Harvard win?

(c) Describe any pattern you see when comparing the length of the series and the percentage that Harvard won.

(d) We will now compare your experimental results to the theoretical probabilities. The theoretical probability of Harvard winning if a single game is played is $\frac{2}{3} \approx 0.67$. To find the theoretical probability that Harvard would win a three-game series, we first need to determine all the different ways that Harvard could win the series. Let H designate a Harvard win and let C designate a Columbia win. One way for Harvard to win the series is if Harvard wins the first two games, HH. The probability of Harvard winning the first two games is $\frac{2}{3} \cdot \frac{2}{3} = \frac{4}{9}$. Altogether, there are three different ways that Harvard could win the series. They are shown below with the probability of each occurring.

$$HH \rightarrow \frac{2}{3} \cdot \frac{2}{3} = \frac{4}{9}$$

$$CHH \rightarrow \frac{1}{3} \cdot \frac{2}{3} \cdot \frac{2}{3} = \frac{4}{27}$$

$$HCH \rightarrow \frac{2}{3} \cdot \frac{1}{3} \cdot \frac{2}{3} = \frac{4}{27}$$

To find the probability that Harvard would win a three-game series, we need to add these three probabilities:

$$\frac{4}{9} + \frac{4}{27} + \frac{4}{27} = \frac{20}{27} \approx 0.74.$$

Thus the theoretical probability that Harvard would win a three-game series is approximately 0.74.

i. Find the theoretical probability that Harvard would win a five-game series. The 10 ways this can be done are listed below.

HHH
$CHHH$ $HCHH$ $HHCH$
$CCHHH$ $CHCHH$ $CHHCH$ $HCCHH$ $HCHCH$ $HHCCH$

ii. Find the theoretical probability that Harvard would win a seven-game series. The 35 ways this can be done are listed below.

$HHHH$
$CHHHH$ $HCHHH$ $HHCHH$ $HHHCH$
$CCHHHH$ $CHCHHH$ $CHHCHH$ $CHHHCH$ $HCCHHH$
$HCHCHH$ $HCHHCH$ $HHCCHH$ $HHCHCH$ $HHHCCH$
$CCCHHHH$ $CCHCHHH$ $CCHHCHH$ $CCHHHCH$ $CHCCHHH$
$CHCHCHH$ $CHCHHCH$ $CHHCCHH$ $CHHCHCH$ $CHHHCCH$
$HCCHHH$ $HCCHCHH$ $HCCHHCH$ $HCHCCHH$ $HCHCHCH$
$HCHHCCH$ $HHCCCHH$ $HHCCHCH$ $HHCHCCH$ $HHHCCCH$

iii. Compare your theoretical probabilities to the probabilities you found using the random number generator. How close are they?

4. **Dice.** In the reading, we showed that the probability of getting at least one 6 in four rolls of a die is approximately 0.52.

   (a) Do you think the probability of rolling at least two 6s in eight rolls is greater than, less than, or equal to 0.52? Explain why.

   (b) Find the theoretical probability of rolling no 6s in eight rolls.

   (c) Find the theoretical probability of rolling exactly one 6 in eight rolls. [Note: You have 8 choices for the die showing a 6.]

   (d) Use your answers to (b) and (c) to compute the theoretical probability of rolling at least two 6s in eight rolls.

5. **Roulette.** We are interested in finding the "best bet" in roulette. Do this by completing the following table to find the expected amount of earnings. Remember that expected value is the same thing as the mean of the probability distribution. The first entry is done for you and the calculation is shown below the table. What is the best bet? Why? [Note: On a roulette table there are 38 numbers. Of these, 18 are red, 18 are black, and 2 are green. Eighteen of these numbers are considered even. A column contains 12 numbers and a row contains 3 numbers.]

Bet	Prob of Losing	Money Not Received	Prob of Winning	Money Received	Expected Value
Red	$\frac{20}{38}$	$-\$1$	$\frac{18}{38}$	$\$1$	$-\$0.05$
Even		$-\$1$		$\$1$	
First 12		$-\$1$		$\$2$	
Column		$-\$1$		$\$2$	
Row		$-\$1$		$\$11$	
Two Adjacent Numbers		$-\$1$		$\$17$	
Single Number		$-\$1$		$\$35$	

The expected value of earnings for the first row was calculated as follows:

- Multiply each probability by the net money received (or lost).
- Add the answers.

$$\frac{20}{38} \cdot (-1) + \frac{18}{38} \cdot 1 = -0.05.$$

6. **Medical Testing.** One of the tests that is available to determine if someone has the human immunodeficiency virus (HIV) is called the Single Use Diagnostic System (SUDS). This test will give a positive result in approximately 97.9% of the cases where someone has the virus and 5.5% of those who do not have the virus.[2] The latter is called a false positive.

---

[2] *Centers for Disease Control and Prevention,* "Rapid HIV Testing: 2003 Update," 28 October 2003, <http://www.cdc.gov/hiv/rapid_testing/materials/ USCA_Branson.pdf>.

Suppose you are living in a city of 100,000 people. It is estimated that approximately 0.3% of the United States population is infected with HIV.[3] We will assume that this percentage is true for the city you are living in as well.

Suppose you test positive for HIV using the SUDS test. Since there is a 5.5% chance of false positives among those without the virus, there is some chance that you do not have the virus, even though you have tested positive. We want to find the probability that you actually have the virus.

(a) Let's first consider those people who have the virus.

    i. What is the estimate for how many people in your city have HIV?

    ii. Suppose all those who have the virus in your city were tested. Approximately how many would test positive and how many would test negative?

    iii. What assumptions are you making in your calculation?

(b) Suppose everyone who did not have the virus were tested.

    i. How many people in your city do not have HIV?

    ii. Of these, how many would you expect to test positive, and how many would test negative?

    iii. What assumptions are you making in your calculation?

(c) Using parts (a) and (b), if everyone in your city were tested for HIV, both those with the virus and those without, how many would test positive? How many would test negative?

(d) You have tested positive for HIV. What is the probability that you actually have the virus? Explain how you arrived at your answer.

(e) Does your answer in part (d) seem surprising? Explain why the probability is so much lower than the 97.9% accuracy for those having the virus and the 94.5% chance that those without the virus test negative.

(f) One erroneous result is a false positive. Another is a false negative. Given the accuracy of the test and the other data given, what is the probability that a person who tests negative for the virus actually has HIV?

---

[3] *Centers for Disease Control and Prevention,* 28 October 2003, "HIV/AIDS Update," <http://www.cdc.gov/nchstp/od/news/At-a-Glance.pdf>.

14

Random Samples

Choosing random samples is a basic principle of good statistical inference. Enforcing randomness eliminates the bias a person might encounter in self-selecting a sample. In this section, we will see what inferential statistics is and why choosing samples properly is important. We will also see what bias and variability are and look at common sampling mistakes.

Inferential Statistics

Populations and Samples

The type of statistics that we looked at earlier in the sections *Displaying Data* and *Describing Data* is called descriptive statistics. These sections gave information about organizing data using frequency distributions, displaying data with histograms and scatterplots, and describing data using means, medians, standard deviations, and regression equations. Inferential statistics, on the other hand, involves making inferences, or decisions, about the entire population when you only have data from a sample. A **population** consists of the entire group of people or objects from which you would like some information. If you wanted information about all the students at a college, for example, the population would consist of all students at the college. A **sample** is the part of the population that you actually examine. For example, to determine information about all the students at a college, your sample might be 100 students who fill out a survey.

Using the entire population to gain information can be very time consuming and expensive. The United States government attempts to do this every ten years when it takes a census. During the census, the goal of the Census Bureau is to contact everyone living in the United States. This process takes a lot of time and money and can never be completely accurate since it is impossible to be sure that you have received information from every person. While there are reasons for taking a census, getting the information desired by the Census Bureau could be accomplished by sampling. Sampling would be less expensive and, if done properly, would produce better results than a census. Not only is sampling more efficient and cost effective than examining the entire population, but there are times when examining the whole population is just not feasible. For example, suppose you were manufacturing light bulbs and wanted to know the average life span for a particular type of bulb. It would not make any sense for you to test all the bulbs you produced by putting them in a lamp and seeing how long they lasted. The only thing you would have to sell is burned-out bulbs! However, taking a random sample of bulbs and finding the average life of the sample would give you a fairly accurate estimate of the average life of all of the bulbs produced. Similarly, Nielson television ratings are based on a sample of about 5000 households. Their viewing habits are then used to estimate the viewing habits of the entire U.S. population.

Just as there is a difference between a population and a sample, there is a difference in the statistical words associated with these. A **parameter** is a number that describes a population. A **statistic** is a number computed from a sample of the population. For example, suppose a nation-wide poll is taken and one of the questions asked is: "Do you favor the formation of a national health care system?" Assume a sample of 1000 people is taken and that 380 of the 1000 respondents favor such a system. The proportion of respondents in the sample that favor the system, $\frac{380}{1000} = 0.38 = 38\%$, is a statistic. It is a numerical measure of a sample property. The corresponding population parameter is the percentage of *all* Americans who favor the formation of a national health care system. Parameters are usually unknown while statistics can be generated through sampling. However, the statistics generated by the sample are typically used to estimate the parameters (i.e., the information about the entire population).

Confidence Interval

If 38% of the people in a sample of 1000 favor the formation of a national health care system, 38% becomes an estimate of the proportion of the entire population that favor such a system. The true population proportion may not be 38%, but hopefully it is something close to 38%. The results of polls often give a "margin of error." For example, it may be reported that 38% of those sampled favor the formation of a national health care system and the margin of error for this poll is plus or minus 3%. This means the poll is saying that the proportion of the population that favor such a system is somewhere between 35% and 41%. In statistics, this type of result is called a confidence interval. A **confidence interval** is an interval estimate computed from sample data such that the interval has a given probability of containing the population parameter. Suppose that the confidence interval 35% to 41% had a 95% probability of containing the population parameter. We would then call this a 95% confidence interval. This means that, using similar polling methods, 95% of the time we would obtain an interval that contains the population parameter.

There are different formulas for different types of confidence intervals. For example, the interval used to estimate a population mean (like the mean life time of light bulbs) is different from that used to estimate a population proportion (like the percentage of the population that favor an issue). The formula used to determine a confidence interval for a population proportion is

$$\hat{p} \pm z \sqrt{\frac{\hat{p}(1 - \hat{p})}{n}}.$$

In this formula, $\hat{p}$ is the sample proportion and n is the sample size.

Since $\hat{p}$ is a number between 0 and 1 and z is a positive number, the formula actually determines two numbers. The first, $\hat{p} - z\sqrt{\hat{p}(1-\hat{p})/n}$, is a number less than $\hat{p}$ and the second, $\hat{p} + z\sqrt{\hat{p}(1-\hat{p})/n}$, is a number greater than $\hat{p}$. The formula

$$\hat{p} \pm z\sqrt{\frac{\hat{p}(1-\hat{p})}{n}}$$

determines an *interval* centered at $\hat{p}$ (known as a confidence interval) by giving the lower and upper endpoints of the interval. The z is called a z-score. The value of z depends on how certain we want to be that our interval contains the population proportion. In other words, the z-score is a number that is dependant on the confidence level. The **confidence level** is the probability that the interval contains the population parameter. The most commonly used confidence level is 95%. This means the interval we obtain has a probability of 0.95 that it will contain the population proportion. For a 95% confidence interval, the z-score used is 1.96. The formula for a 95% confidence interval for a population proportion is

$$\hat{p} \pm 1.96\sqrt{\frac{\hat{p}(1-\hat{p})}{n}}.$$

If we want an interval that is more likely to contain the population proportion, the interval must be wider. This means that our z-score must be larger. For example, the z-score is 2.576 for a 99% confidence interval. (See Figure 14.1.) For the rest of this section, we will be working exclusively with 95% confidence intervals.

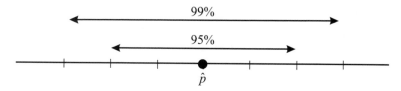

FIGURE 14.1
A 99% confidence interval is wider than a 95% confidence interval because it is more likely to contain the population proportion.

Example 1. In 1997, a poll of 150 students was taken at midwestern college to determine if students favored a proposal that would prohibit smoking in all residence halls. The poll showed that 93 of the 150 students favored such a ban. While the 93/150 = 0.62 shows that 62% of the polled students favor such a ban, what does this say about all the students at the college? Determine a 95% confidence interval to answer this question.

Solution. We know that $\hat{p} = 0.62$ and $n = 150$. Putting these numbers into our formula, we get

$$0.62 \pm 1.96\sqrt{\frac{0.62(1-0.62)}{n}}$$

$$0.62 \pm 1.96\sqrt{\frac{0.62(0.38)}{150}}$$

$$0.62 \pm 0.078.$$

Since $0.62 - 0.078 = 0.542$ and $0.62 + 0.078 = 0.698$, we can be 95% confident that the proportion of all students at the college that favor a ban on smoking in residence halls is somewhere between 0.542 and 0.698 or between 54.2% and 69.8%. ■

It was stated earlier that a sample of 1000 produced a margin of error of plus or minus 3%. Let's see where this 3% comes from. In Example 1, the margin of error was 0.078 or 7.8%. In general, for a 95% confidence interval, the margin of error is $1.96\sqrt{\hat{p}(1-\hat{p})/n}$. You can see that this is dependent on $\hat{p}$ and n. If we want a margin of error of plus or minus 3% before we run a poll, how do we know what $\hat{p}$ is? The answer is we do not. The key is to take a larger sample, which will reduce the margin of error. To do so, we need to find a sample size that is large enough for any value of $\hat{p}$. All we need to do is find the largest possible value for $\hat{p}(1-\hat{p})$. Table 14.1 shows values of $\hat{p}(1-\hat{p})$ for various values of $\hat{p}$.

TABLE 14.1

Values of $\hat{p}(1-\hat{p})$ for various values of $\hat{p}$. Notice that the maximum value for $\hat{p}(1-\hat{p})$ is 0.25.

$\hat{p}$	0.1	0.2	0.3	0.4	0.5	0.6	0.7	0.8	0.9
$\hat{p}(1-\hat{p})$	0.09	0.16	0.21	0.24	0.25	0.24	0.21	0.19	0.09

Notice that the maximum value for $\hat{p}(1-\hat{p})$ is 0.25, which occurs when $\hat{p} = 0.5$. Therefore, to find the sample size needed so the margin of error is no larger than plus or minus 3%, we need to solve the equation $1.96\sqrt{\frac{0.25}{n}} = 0.03$. Doing so gives a value of approximately 1067. This means that if we take a sample size of 1067, we are guaranteed a margin of error of no more than 3%. Many national polls typically choose sample sizes of about 1000. If the results of the polls need to be more precise, this sample size is increased. Increasing the sample size will reduce the margin of error; it also involves more work and often becomes more costly.

Collecting a Sample

Random Samples

Collecting a proper sample is key to producing good inferential statistics. Without it, the conclusions that you draw from your sample are usually worthless. Typically, national polls are conducted by sampling approximately 1000 people since this gives a margin of error of no more than 3%. From this relatively small number, the opinions of more than 200 million Americans can be inferred with a relatively high level of confidence.

For a sample to be properly collected, it must be random. The most basic sampling technique is a simple random sample. A **simple random sample** of size n consists of n units taken from a population in such a way that every set of n units has an equal probability of being in that sample. For example, suppose a sample of size 12 was taken from the 3000 students at a college. In order for this sample to be random, every possible group of 12 students must have an equal chance of being the sample. The key here is randomness. This is very important. We do not want to just use the first 12 students we see. For example, if we were doing the survey on smoking in residence halls and see 12 people standing out in front of the library and ask them their opinions, we might have just asked 12 smokers what they thought about smoking in residence halls. We can be pretty sure that smokers opinions on this issue differ from those of nonsmokers and our results would not be very accurate.

There are other types of sampling techniques that can also produce results that effectively represent the entire population. One of these is called a stratified random sample. In a **stratified random sample**, the population is first divided into different groups and then simple random samples are taken from each group. For example, the students at a college could be divided into different groups according to whether they are freshmen, sophomores, juniors, or seniors. After this grouping, a simple random sample of size three could be taken from each of the four class levels. The twelve students chosen would represent a stratified random sample of college students. Note that this type of sample is different from a simple random sample. In a simple random sample, a specific group of twelve seniors has an equal probability of being selected as a specific group consisting of three students from each class. In the stratified sample, it is impossible to select a group of 12 seniors.

Bias and Variability

When selecting a sample to determine information about the entire population, there are two things that should be minimized: bias and variability. **Bias** exists when the results obtained from an experiment are in some way prejudiced. Bias often occurs when the sample selected is not representative of the entire population. **Variability** exists when the results you obtain vary significantly when you use a different sample. These two ideas are illustrated in Figure 14.2. Assume the center of the target is the parameter, p, for the entire population and the dots are the different $\hat{p}$ determined by several different samples. If variability is low, the dots will be clustered together. If bias is low, the dots will be around the center.

If a sample is selected randomly, the possibility of getting a biased sample that systematically favors a particular outcome is greatly reduced. There are, however, other

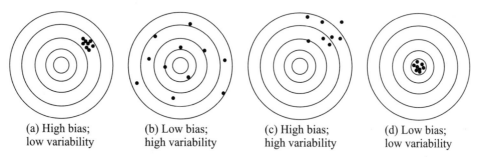

(a) High bias;
low variability
    (b) Low bias;
high variability
    (c) High bias;
high variability
    (d) Low bias;
low variability

FIGURE 14.2
An illustration of showing varying levels of variability and bias.

things that can bias a study. For example, the way in which questions are worded can bias the results. Think about how people might respond when asked, "Do you favor increased trade with the Communist government of the People's Republic of China?" compared with, "Do you favor increased trade with the people in China that could result in greatly benefiting the economy in the United States?" These two questions would likely produce different results. Even subtle changes in wording questions, such as changing "welfare" to "aid to the poor" can influence the responses people give to a question.

To keep variability low, a large sample size is usually needed. However, large does not mean hundreds of thousands. If a random sample of about 1000 people is used in a study, the results might include a statement such as "margin of error of plus or minus 3%." This margin of error is usually acceptable. If the sample were to be reduced to 100 instead of 1000, the margin of error would increase to plus or minus 10% which is too high to be acceptable for most studies. So bias is generally controlled with good sampling design that includes randomness and careful design of questions, while variability is generally controlled with sample size.

Examples of Poor Sampling Methods

Current statistical studies, as well as those in the past, do not always use proper sample collecting techniques. A classic example of this occurred in 1936. *The Literary Digest*, a popular magazine of its time, predicted that Alf Landon would win a landslide victory over Franklin Roosevelt in the presidential election. Since you have never heard of President Landon, where did the *Literary Digest* poll go wrong? The magazine sent out about ten million ballots and over two million of them were returned. This is a huge sample size. Remember that a national sample of about 1000 can give excellent results. So the size of the sample was not a problem. The problem occurred in how the sample was selected. The names of the people that were sent ballots came from lists of people who owned telephones and automobiles. Remember that 1936 was in the middle of the Great Depression so owning a telephone or automobile was quite a luxury. Wealthier people tended to belong to the Republican party. Therefore, this sample was heavily biased towards the Republican candidate for president, Alf Landon.

A more recent example of poor sampling techniques was a study done by Shere Hite as the basis for her book, *Women and Love, a Cultural Revolution in Progress*, published in 1987. In this book, Hite reports that women, for the most part, are not happy in their relationships with men and that men are to blame for this. Some of her startling findings included:

- 95% of women report being emotionally and psychologically abused by the men they love.

- 70% of women married five years or more report having extramarital affairs.

- 87% of the married women said they have their deepest emotional relationship with a woman friend.

These results are biased because of the sampling technique used by Hite. She mailed 100,000 questionnaires to women who were members of various women's groups. The questionnaire consisted of 127 essay questions. Only 4.5% of the surveys were returned.

The procedure used by Hite did not produce a random sample for many reasons. First of all, she only sampled from women who join women's groups instead of a random sample chosen from a larger population of women. Secondly, she received results only from those who took the time to complete a long survey. Thus, the results are highly biased. Even though the statistical methods used may be questionable, she sold many books and respectable magazines, such as *Time*, reported her results.

One conclusion from these two examples is "do not believe everything you read." Examine how the statistical results were obtained before giving credence to the results. Many polls lack randomness and thus can be highly biased. You have probably seen polls conducted by television shows requesting people to call in their opinion or a magazine article requesting people to complete a survey. The people that respond to these types of surveys are self-selecting themselves into the sample. This is not a random sample and the results will probably not reflect the opinions of the population from which the sample was taken.

Application: Religion Polls

One question asked by the Gallup polling organization concerns the religious beliefs of Americans. One of the questions used is given in Example 2.

Example 2.   One of the questions that the Gallup polling organization often uses to assess how committed Americans are to religious organizations is,

> "Did you, yourself, happen to attend church or synagogue in the last seven days, or not?"

In December of 2002, 43% of 1009 people responded that yes they did attend a church or synagogue in the last seven days. Find a 95% confidence interval for the proportion of all Americans that attended church or synagogue in the last seven days. Also, find the sample size needed to have a margin of error of 1%.

Solution.   Using the formula $\hat{p} \pm 1.96 \sqrt{\dfrac{\hat{p}(1 - \hat{p})}{n}}$, where $\hat{p} = 0.43$ and $n = 1009$, we get the following.

$$0.43 \pm 1.96 \sqrt{\frac{0.43(1 - 0.43)}{1009}}$$

$$0.43 \pm 1.96 \sqrt{\frac{0.43(0.57)}{1009}}$$

$$0.43 \pm 0.03.$$

Therefore, we can be 95% confident that the proportion of all Americans that attended church or synagogue in the last seven days is between 40% and 46%.

To find the sample size needed to get a margin of error of 0.01 we need to solve the equation

$$1.96 \sqrt{\frac{0.25}{n}} = 0.01$$

for n. We start out by dividing both sides by 1.96 and then squaring both sides.

$$1.96 \sqrt{\frac{0.25}{n}} = 0.01$$

$$\sqrt{\frac{0.25}{n}} = \frac{0.01}{1.96}$$

$$\frac{0.25}{n} = \left(\frac{0.01}{1.96}\right)^2.$$

Now we invert both sides and then multiply both sides by 0.25.

$$\frac{n}{0.25} = \left(\frac{1.96}{0.01}\right)^2$$

$$n = \left(\frac{1.96}{0.01}\right)^2 0.25$$

$$n = 9604.$$

Therefore, to have a margin of error of 1% we would need a sample size of 9604. In this example, we used 0.25 for $\hat{p}(1 - \hat{p})$. This is what you would do if you had no idea what your sample proportion might be. In this example, however, we did have an idea of what our sample proportion would have been from the earlier poll, $\hat{p} = 43\% = 0.43$, so $1 - \hat{p} = 0.57$. We could have used $0.43 \times 0.57 = 0.2451$ instead of 0.25. This would have given us a slightly smaller sample size. ■

The question asked by Gallup about church attendance is designed to find out what proportion of Americans attend church during a given week. Notice that the question is not stated, "Do you usually attend church?" or "Are you a regular church goer." These types of questions can be answered "yes" by those who have good intentions of going to church or think of themselves as church attenders, but rarely do.

Reading Questions for Random Samples

1. Explain how a population and sample differ.

2. Explain how a parameter and a statistic differ.

3. Suppose a poll finds that 62% of the respondents favor a ban on gambling casinos in the state and the poll has a margin of error of plus or minus 3%. What does this mean?

4. When do you use the formula $\hat{p} \pm z\sqrt{\frac{\hat{p}(1 - \hat{p})}{n}}$? What do $\hat{p}$, z, and n represent?

5. A poll of 200 people in Holland, Michigan was taken to determine if Internet filters should be required on library computers. The poll showed that 147 people favored Internet filters.

(a) Give a 95% confidence interval for the proportion of all the people in Holland that favor Internet filters.

(b) Does your confidence interval from part (a) have a margin of error of 3%? Why or why not?

6. The reading claims that you solve $1.96\sqrt{\frac{0.25}{n}} = 0.03$ to find the sample size needed for a margin of error of 3%.

   (a) Why does this equation only use the $z\sqrt{\frac{\hat{p}(1-\hat{p})}{n}}$ part of the formula instead of using $\hat{p} \pm z\sqrt{\frac{\hat{p}(1-\hat{p})}{n}}$?

   (b) Explain why $n = \left(\dfrac{1.96\sqrt{0.25}}{0.03}\right)^2$.

   (c) Explain why the reading says that n is approximately 1067. Wouldn't it be more accurate to say n is approximately 1067.111?

7. What is the difference between a simple random sample and a stratified random sample?

8. If you repeat a sample several times and get very similar answers, you have (low, high, can't tell) variability and (low, high, can't tell) bias.

9. Describe ways to keep bias low when conducting a survey.

10. Describe how to keep variability low when conducting a survey.

11. Why did the *The Literary Digest* poll of the 1936 presidential election produce such biased results?

12. List some reasons that caused the statistical work done by Shere Hite to produce biased results.

13. Suppose a poll is conducted by asking you to call 1-888-555-DEMO if you think a Democrat should be the next president and 1-888-555-REPB if you think a Republican should be the next president. Why might this method produce biased results?

14. In May of 2003, Annika Sorenstam competed in a PGA Tour golf tournament, the first female to do so in 50 years. The Gallup Organization conducted a poll at that time asking, "As a general rule, would you like to see more women compete with men in professional golf tournaments, or not?" Out of 391 golf fans asked this question, 59% answered yes, they would like to see more women compete with the men.[1]

   (a) Find a 95% confidence interval for all American golf fans that would like to see more women compete with the men in professional golf tournaments.

   (b) Find the sample size needed to have a margin of error of no more than 0.02.

---

[1] Gillespie, Mark, "Gallup Poll Analysis - Americans Heartily Endorse Annika Sorenstam's Appearance on PGA Tour," May 23, 2003, (visited 2 June 2003), <http://www.gallup.com/poll/releases/pr030523b.asp>.

Random Samples: Activities and Class Exercises

1. **Population.** To determine the size of a population, such as fish in a lake, it is impractical or impossible to count all of them. To estimate the size of the population indirectly, biologists often capture a sample of the population, mark that sample in some way and release the sample. Later, another sample is captured (or many samples over a period of time) and the portion that is marked is used to estimate the size of the entire population. The simplest form of this type of mark and recapture techniques is a single marking and single recapture on a closed population (i.e. one in which no immigration, emigration, or recruitment occurs). With this technique, a portion of the population is captured, marked and released. A sufficient amount of time is allowed for the sample to disperse throughout the entire population and then another sample is taken. In the second sample, the ratio of marked animals, m_2, to the sample size, n_2, should be similar to the ratio of the original number marked in the first sample, n_1, to the entire population, N. That is,

$$\frac{m_2}{n_2} \approx \frac{n_1}{N}.$$

Solving this for N, we get

$$N \approx \frac{n_1 n_2}{m_2}.$$

If we let $\hat{N}$ be an estimate of the size of the population, then the formula we can use to find $\hat{N}$ is

$$\hat{N} = \frac{n_1 n_2}{m_2}.$$

Suppose we want to estimate the number of trout in a lake. We will simulate this method by estimating the number of bingo chips in a bag.

   (a) Using a bag of red bingo chips and a cup, take a cupful of the red bingo chips from the bag as a sample. Count them and "mark" this sample by replacing the red bingo chips with blue bingo chips. The number of blue bingo chips is n_1. Record your value for n_1.

   (b) Mix the blue bingo chips back with the rest of the red chips in the bag. Take a second sample, another cupful and determine n_2 and m_2. Record your values for n_2 and m_2.

   (c) Use your values for n_1, n_2, and m_2 to determine $\hat{N}$. This is a point estimate for the population size (or all the bingo chips in the bag).

   (d) Compare your estimate for $\hat{N}$ with the estimates obtained by the rest of the class.

2. **Boxes.** The following page contains 80 squares of size 1, 4, 9, 16, 25, and 36. We are interested in finding the average (mean) size square on the page. Have each person in your group estimate the mean in the following ways. Record each individual result. They will be combined with those of the entire class so you can answer the following questions.

- Just by looking at the squares, guess the mean.

- Choose five squares whose sizes you think are representative of all of them. Find the mean for your sample.

- Randomly choose five squares by choosing five numbers at random. Use a table of random numbers or the random number function on a calculator to choose random numbers between 1 and 80.

- Randomly choose ten squares by the same methods as you used to randomly choose five squares.

(a) Which method seems to have the most bias? Explain.

(b) Which method seems to have the most variability? Explain.

(c) Which method is best for estimating the mean size square?

3. **Proportion.** Polls are often taken to estimate a proportion of a population that will vote for a certain candidate, believe a certain way about an issue, or have a certain characteristic. We want to model how polling reflects the population by performing an experiment using beads. Using a bag of clear beads and red beads we want to estimate the proportion of red beads in the bag.

(a) Take a random sample of 20 beads from the bag. What is your sample proportion, $\hat{p}$? To find $\hat{p}$, count the number of red beads and divide it by the total number of beads.

(b) Determine a 95% confidence interval for the population proportion of all red beads in the bag.

(c) Repeat parts (a) and (b), this time taking random samples of 100 beads.

(d) Is the width of your confidence interval wider or narrower when you used a larger sample? Explain why.

4. **Time Management.** We are going to look at how college students spend their time, by comparing results you come up with to those found at a midwestern college.

(a) Every student in the class should estimate how long he or she spends on each activity during a typical weekday.

- Hours Studying
- Hours Working
- Hours Sleeping
- Hours Eating
- Hours Watching TV

Compile the class data on the board. Find the mean for each category.

(b) For one week and two weekends in November of 1993, 175 students from a midwestern liberal arts college kept time logs of their activities. To get a representative sample of students, the population was stratified by class and gender and from these the students were then randomly chosen. The time logs were then compiled by students working in the college's research center and average time spent on various activities was calculated. From the results it was found that on weekdays these students spent on the average:

- 4.25 hours studying
- 0.98 hours working

- 7.56 hours sleeping
- 1.54 hours eating
- 0.72 hours watching TV

i. Describe how your class results compare with the research findings given above.

ii. Describe ways in which the research center survey was conducted to make it a better method for estimating average times that students spend on various activities than the method you used.

iii. Do you think that gathering information the way that you did includes more variability than the research center methods? Explain why or why not.

iv. Do you think that gathering information the way that you did includes more bias than the research center methods? Explain why or why not.

5. **Recycling.** Two articles, "Environmental troopers" and "Poll says Americans want less wasteful packaging in stores," came out in April of 1991 within a few days of each other. They appeared in *USA Weekend* and *Grand Rapids Press*, respectively. The *USA Weekend* article contained the following.

> Reilly and Merrow (administrator of the EPA and Sierra Club president) are encouraged by the results of USA WEEKEND's special survey on environmental gripes and habits. More than 70 percent of the 11,000 readers who answered the "Speak Out" survey by mail earlier this year say they will do anything—even pay more—to help the environment. ... What do you recycle now: Newspaper, 84%; Glass, 76%; Aluminum, 88%; Plastic, 67%; All four, 65%.

The *Grand Rapids Press* article contained the following.

> The poll found a majority of people say they recycle: 58 percent say they recycle cans regularly, but that drops off to 42 percent for newspapers and 36 percent for bottles. ... The telephone survey of 1001 adults took place on a Friday through Tuesday and was conducted by ICR Survey Research Group of Media, PA. Results have a margin of sampling error of plus or minus three percentage points.

(a) Notice that one article says that 84% of the respondents recycle newspapers while the other says only 42% of the respondents recycle newspapers. Which article gives a more accurate representation of the percentage of Americans that recycle newspapers?

(b) Describe the flaws in the survey methods of the article that gives the worst results. Do these results come from a problem with variability or bias?

Appendix: Instructions for the TI-83 Graphing Calculator

Graphing a function

You can use your TI-83 to construct an xy-plot of almost any function. This process will be illustrated by graphing the function $y = 2x + 3$ and the function $y = x^2$. (This information is in Chapter 3 of the TI-83 Guidebook.)

1. The first step in graphing a function is to define the function so that the calculator knows what you want to graph. Press $\boxed{\text{MODE}}$ and select **Func** on the fourth line down (if it is not already selected). You can do this by using the arrow keys. When the cursor is on top of the word **Func**, press the $\boxed{\text{ENTER}}$ key. Your calculator display should look like the one shown below. If not, use $\boxed{\triangle}$ or $\boxed{\triangledown}$ to get to the correct line and use $\boxed{\triangleright}$ or $\boxed{\triangleleft}$ to highlight the correct word. Once you are in the correct mode for graphing, press the $\boxed{\text{QUIT}}$ key (this is $\boxed{\text{2nd}}$ followed by $\boxed{\text{MODE}}$).

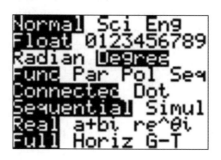

2. Press the $\boxed{\text{Y=}}$ button (the leftmost blue button directly under the screen). This is the screen where you enter the function you wish to graph. If any of the **Plot1**, **Plot2**, or **Plot3** items are highlighted, clear these by using the arrow keys to place the cursor over the appropriate word and pressing the $\boxed{\text{ENTER}}$ key. If there are any functions already defined, place the cursor to the immediate right of the equal sign and press the $\boxed{\text{CLEAR}}$ button. Your screen should now be empty and resemble the following.

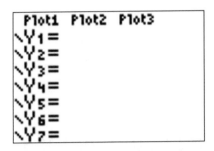

3. Enter the function $y = 2x + 3$ into your calculator by moving the cursor to the right of the equal sign for **Y1** and pressing $\boxed{2}$ $\boxed{X,T,\theta,n}$ $\boxed{+}$ $\boxed{3}$.

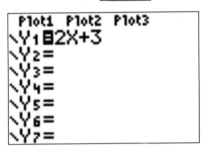

4. We also need to set an appropriate window for viewing the graph. Press the $\boxed{\text{WINDOW}}$ button (the second blue button from the left). **Xmin** is the smallest input for x that will be displayed on the graph, **Xmax** is the largest input for x that will be displayed on the graph, **Xscl** defines the distance between tick marks on the horizontal axis. **Ymin** is the smallest output for y that will be displayed on the graph, **Ymax** is the largest output for y that will be displayed on the graph, **Yscl** defines the distance between tick marks on the vertical axis, and **Xres** tells the calculator at which pixels it should evaluate the function in order to draw the graph. **Xres = 1** means the calculator will evaluate the function at every pixel. For now, leave all of window items defined with their default values. (These are shown in the following screen. These can also be obtained by pressing $\boxed{\text{ZOOM}}$ $\boxed{6}$.) Press the $\boxed{\text{GRAPH}}$ key (blue button furthest to the right) to display the graph.

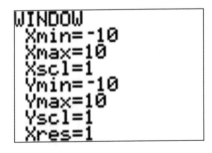 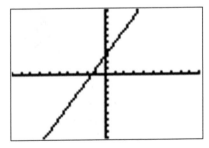

5. Now enter the second graph, $y = x^2$, by pressing the $\boxed{\text{Y=}}$ button, moving the cursor to Y2, pressing $\boxed{X, T, \theta, n}$ and $\boxed{x^2}$. Press $\boxed{\text{GRAPH}}$ to display both graphs.

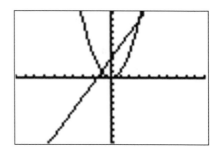

6. We are interested in finding the coordinates of the left-most point where the graphs intersect. To do this, we first want to zoom in closer to the point. Press the $\boxed{\text{ZOOM}}$ key (the blue button in the middle) and select **2: Zoom In** by using the arrow keys to highlight your selection and pressing $\boxed{\text{ENTER}}$. This will return you to your graph. Place the cursor (using the arrow keys) over the point of intersection and press $\boxed{\text{ENTER}}$. The calculator will re-draw the graph using the point where you placed the cursor as the center of the screen. If you desire, repeat the process to zoom in even closer.

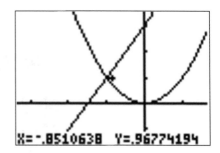

7. To find the coordinates of the intersection point, press $\boxed{\text{2nd}}$ then $\boxed{\text{TRACE}}$. This selects the $\boxed{\text{CALC}}$ function. Choose the **5:intersect** option. The calculator will ask for the first curve. The curser should be on the line and you need to just hit enter. It will then ask for the second curve. The curser should now be on the parabola and again hit enter. It will finally ask for a guess. Just use the arrow keys to put the curser approximately at the intersection point and hit enter. You should now see that the coordinates of the intersection are $(-1, 1)$.

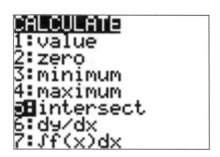

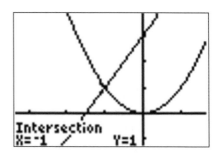

Finding x- and y-intercepts

We can use the calculator to find the x- and y-intercepts of a function given symbolically. We will demonstrate this with the functions $y = x^3 + 2x^2 - 7x - 6$.

1. Enter the function in your calculator by pressing $\boxed{\text{Y=}}$ $\boxed{\text{X}}\boxed{\wedge}\boxed{3}\boxed{+}\boxed{2}\boxed{\text{X}}\boxed{\wedge}\boxed{2}\boxed{-}$ $\boxed{7}\boxed{\text{X}}\boxed{-}\boxed{6}$. Graph it using the viewing window **Xmin**=-5, **Xmax**=5, **Ymin**=-10, **Ymax**=10.

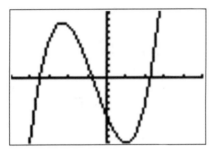

The y-intercept is where the function crosses the y-axis. Since this occurs when $x = 0$, we can see by substituting zero for x that $y = 0^3 + 2 \times 0^2 - 7 \times 0 - 6 = -6$. But as a check, let us use the calculator to find the y-intercept.

2. Press $\boxed{\text{2nd}}$ then $\boxed{\text{TRACE}}$. This selects the $\boxed{\text{CALC}}$ function.

The first option on the list is **1:Value**. This calculates the output or y-value for any given input or x-value. Select this option and, when asked to give a x-value, enter zero.

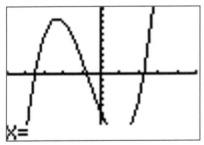   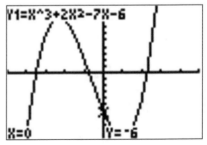

Notice that the y-value is -6. You can use this option to find the corresponding y-value (output) for any x-value (input). If you are in the $\boxed{\text{TRACE}}$ mode, you can do the same thing by entering an x-value at any time.

3. We now want to calculate the x-intercepts. These occur where the graph of the function crosses (or touches) the x-axis. These are also called the *zeros* or *roots* of the function because $y = 0$ when a function touches the x-axis. So the x-intercepts are the values of x such that $0 = x^3 + 2x^2 - 7x - 6$. We will illustrate this by finding the coordinates for the leftmost x-intercept. Press the $\boxed{\text{2nd}}$ and $\boxed{\text{TRACE}}$ keys to select the $\boxed{\text{CALC}}$ option. Chose the **2:zero** option and press $\boxed{\text{ENTER}}$. [Note: you may have to use $\boxed{\triangle}$ and $\boxed{\triangledown}$ to select the correct function.] Once selected, the calculator asks you for the left-bound.

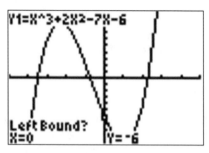

This is asking you to give a lower (left) bound for the zero you are seeking. Use the $\boxed{\triangleleft}$ and $\boxed{\triangleright}$ keys to move the cursor to the left of the leftmost zero and press $\boxed{\text{ENTER}}$.

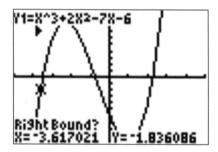

The calculator now asks you for the right bound. Use the $\boxed{\triangleright}$ to move the cursor on the function slightly to the right. Press $\boxed{\text{ENTER}}$.

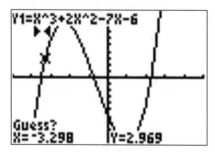

The calculator now asks for a guess. The calculator is asking you to move the cursor close to the x-intercept. Move the cursor close to the leftmost point where the graph crosses the x-axis and press $\boxed{\text{ENTER}}$.

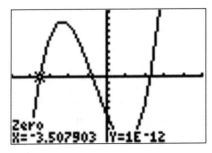

The coordinates of the x-intercept are shown at the bottom of the screen. [Note: the y value may be something very, very small like 1E-12 instead of zero. 1E-12 is the number $1 \times 10^{-12} = 0.000000000001$.]

Constructing a Histogram

You can use your TI-83 to construct a histogram of the height data found in Table 4.2 on p. 68. All 50 pieces of data can either be entered into one list or the data can be entered in the form of a frequency distribution. We will illustrate the second method since it requires fewer keystrokes. To do this, we will input the first column of Table 4.2 (the heights) into **L1** and the second column (the frequencies) into **L2**. (This information can be found on pages 12-2 and 12-32 of the TI-83 Guidebook.)

1. To open the statistical list editor, press ⌐STAT⌐ and then select **1:Edit** from the menu. Press ⌐ENTER⌐. Your cursor should now be at the first entry in **L1**.

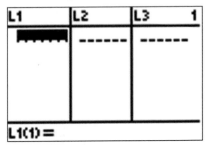

2. If **L1** and **L2** are empty, you can input the data. (If **L1** and **L2** are not empty, they need to be cleared first. To do this, press ⌐△⌐ ⌐CLEAR⌐ ⌐ENTER⌐.) Enter the data by inputting ⌐6⌐ ⌐0⌐ ⌐ENTER⌐ ⌐6⌐ ⌐1⌐ ⌐ENTER⌐, and so on. In a similar fashion, input the frequencies into **L2**. To get the cursor from **L1** to **L2**, press ⌐▷⌐.

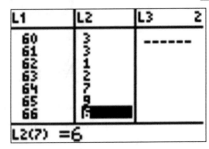

3. To graph the histogram, you need to go to the **STAT PLOT** menu. To do this, press ⌈2nd⌉ ⌈Y=⌉. Now select **Plot 1** by pressing ⌈ENTER⌉. On this menu, we want the plot turned **On**, and the **Type** to be a histogram. (To do this, use the arrow keys and the ⌈ENTER⌉ button.) We also want the **XList** to be **L1** and the **Freq:** to be **L2**. (**L1** is ⌈2nd⌉ ⌈1⌉ and **L2** is ⌈2nd⌉ ⌈2⌉.)

4. We need to set an appropriate window to view our histogram. First, make sure all other graphs and statistical plots are turned off. The easiest way to get a window that is *close* to what we want is to press ⌈ZOOM⌉ ⌈9⌉. This is the **ZoomStat** feature. You should now see a histogram on your screen.

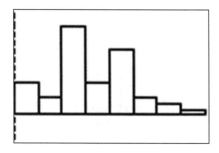

5. This default will not group our data into integer groupings (i.e., $60, 61, 62, \ldots, 72$). To do this, press ⌈WINDOW⌉ ⌈▽⌉ ⌈▽⌉ ⌈1⌉. This makes **Xscl=1** and causes the x-axis to be divided into integer increments. Also, since the largest frequency is 9, our window needs to be a little larger. Therefore, use the arrow keys and set **Ymax=10**. Press ⌈GRAPH⌉ to view the new histogram. Notice that the TI-83 does not put labels on the axes.

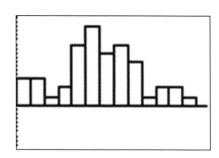

Constructing a Scatterplot

To use your TI-83 to construct scatterplots, you first need to enter your two-variable data into two lists. We demonstrate this by constructing a scatterplot of the height and arm span data found in Table 4.3 on p. 69. (This information can be found on page 12-3 of the TI-83 Guidebook.)

1. Using the height and arm span data from Table 4.3, enter the height data in **L1** and the arm span data in **L2** using the same method described earlier in this appendix for entering the data for the histogram.

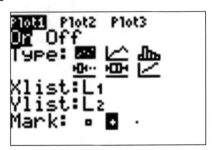

2. To set up the graph for the scatterplot, first go to the **STAT PLOT** menu. On this menu select **Plot 1** and make sure the plot is turned **On**. The **Type** should be a scatterplot (the first one listed), the **XList** should be L1, $\boxed{\text{2nd}}$ $\boxed{1}$, and the **YList** should be **L2**, $\boxed{\text{2nd}}$ $\boxed{2}$. We have chosen the **Mark** to be the +.

3. To set an appropriate window to view the scatterplot, press $\boxed{\text{ZOOM}}$ $\boxed{9}$. This is the **ZoomStat** feature. You should now see a scatterplot on your screen. (Note: You should make sure all other graphs and statistical plots are turned off.)

Constructing an xy-line

To use your TI-83 to construct an xy-line, you first need to enter your two-variable data into two lists. We will demonstrate this by constructing an xy-line of the natural gas data found in the following table. (This information can be found on page 12-31 of the TI-83 Guidebook.)

1. The cost of natural gas per month is given in the following table. To make an xy-line of these data, input the month data in **L1** and the cost of the gas in **L2**. Make sure any previous data is deleted.

Month	Gas	Month	Gas
1	$27.73	13	$32.15
2	$19.73	14	$16.48
3	$11.30	15	$12.92
4	$11.76	16	$12.42
5	$12.81	17	$12.92
6	$23.96	18	$15.49
7	$34.16	19	$29.34
8	$50.85	20	$57.57
9	$75.87	21	$58.15
10	$75.29	22	$59.62
11	$72.73	23	$53.95
12	$45.44	24	$43.60

```
L1          L2          L3      1
┌──────┐    27.73     ------
│1     │    19.73
2           11.3
3           11.76
4           12.81
5           23.96
6           34.16
7
L1(1)=1
```

2. To set up the graph for the xy-line, go to the **STAT PLOT** menu. On this menu select **Plot 1** and make sure the plot is turned **On**. The **Type** should be an xy-line (the second one listed), the **XList** should be **L1** and the **YList** should be **L2**. We have chosen the **Mark** to be the +.

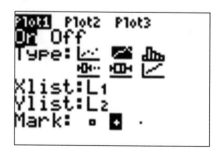

```
Plot1  Plot2  Plot3
On  Off
Type: L⋯  ⬛  dln
      ⊶  ⊶  ⬋
Xlist:L1
Ylist:L2
Mark:  □  ■  .
```

3. To set an appropriate window to view our xy-line, press $\boxed{\text{ZOOM}}$ $\boxed{9}$. This is the **ZoomStat** feature. You should now see the xy-line on your screen. (Note: You should again make sure all other graphs and statistical plots are turned off.)

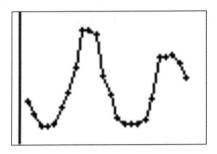

Computing One Variable Statistics

To use your TI-83 to compute mean, median, and standard deviation, you first need to enter your data into a list. We will demonstrate this with the following set of numbers:

$$3, 6, 9, 2, 3, 6, 7, 8.$$

1. To open up the statistical editor, press $\boxed{\text{STAT}}$ and select **1:Edit** from the menu. Press $\boxed{\text{ENTER}}$. Your cursor should now be at the first entry in **L1**. If **L1** is empty, you can just proceed to input the data. (If **L1** is not empty, it needs to be cleared first. To do this, press $\boxed{\triangle}$ $\boxed{\text{CLEAR}}$ $\boxed{\text{ENTER}}$.)

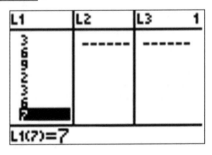

2. Once the data is entered, press $\boxed{\text{STAT}}$ and move the cursor over to the **CALC** menu by pressing $\boxed{\triangleright}$. Press $\boxed{\text{ENTER}}$ to set up the calculator for finding statistics for 1-variable data.

3. To get the calculator to calculate values for **L1**, you need to input **L1** after **1-Var Stats**. Do this by pressing $\boxed{\text{2nd}}$ $\boxed{1}$.[2] Now pressing $\boxed{\text{ENTER}}$ will give you a list of statistics. The first item on your list, $\bar{x} = 5.5$, is the mean, the fifth item on your list, $\sigma x = 2.397915762$ is the standard deviation, and if you scroll down by pressing $\boxed{\triangledown}$ you will find, $Med = 6$, which is the median. You can also find values for Q_1 (the first quartile) and Q_3 (the third quartile).

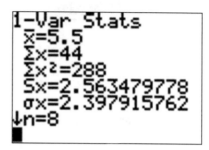

The meaning of each of the 1-Var Stats are:

$\bar{x}$	mean
$\sum x$	sum of the data points
$\sum x^2$	sum of the squares of the data points
Sx	standard deviation for a sample
σx	standard deviation
n	number of items in your data set
$minX$	smallest value in your data set
Q_1	first quartile
Med	median
Q_3	third quartile
$maxX$	largest value in your data set

[Note: The standard deviation we defined in this book, σx, is for a population rather than a sample. When calculating the standard deviation for a sample, Sx, you divide the sum of the squared differences by $n - 1$ rather than by n.]

4. You can also use your calculator to find these same statistics if the data are given in a frequency distribution. Suppose we wanted to find the mean of the following data representing heights, in inches, of a group of students.

Height	66	67	68	69	70	71	72
Frequency	2	4	7	8	6	2	4

To do this, put the heights in **L1** and the frequencies in **L2**. Once the data is entered, press $\boxed{\text{STAT}}$ and move the cursor over to the **CALC** menu by pressing $\boxed{\triangleright}$ and then $\boxed{\text{ENTER}}$. To get the calculator to calculate values for **L1** and **L2** you need to input **L1,L2** after **1-Var Stats**. Do this by pressing $\boxed{\text{2nd}}$ $\boxed{1}$ $\boxed{,}$ $\boxed{\text{2nd}}$ $\boxed{2}$. Now pressing $\boxed{\text{ENTER}}$ will give you a list of statistics.

---

[2] The calculators default list is **L1**, so if your list of numbers is in **L1**, you do not really need to enter **L1**. However if it is any other list, you must identify that particular list.

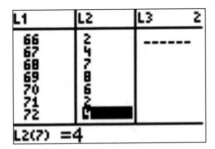

Computing the Regression Equation and the Correlation

To use your TI-83 to compute the linear regression equation and the correlation, you first need to enter your data into two lists. We will demonstrate this with the data from the following table.

Height	152	160	165	168	173	173	180	183
Arm Span	159	160	163	164	170	176	175	188

1. Before you enter the data in the calculator, you need to make sure it is set in the right mode to calculate correlation.[3] This is done by turning the "diagnostic" on. To do this press 2nd 0. This is the catalog. It is a list of calculator commands in alphabetical order. Toggle down to **DiagnosticOn** by holding down ▽. Then press ENTER ENTER and the calculator will now be in the proper mode to calculate correlation.

2. To open up the statistical editor, press STAT and select **1:Edit** from the menu by pressing ENTER. Input the height data in **L1** and the arm span data in **L2**. If necessary, first clear lists **L1** and **L2** by using the arrow keys to scroll the cursor to the top of the list and then press CLEAR ENTER.

---

3 This step is specifically for the TI-83. If you are using a TI-82, skip to number 2.

3. Once the data is entered, press $\boxed{\text{STAT}}$ and move the cursor over to the **CALC** menu by pressing $\boxed{\triangleright}$. Now press $\boxed{\triangledown}$ three times so that the cursor is on **4:LinReg(ax+b)**.

```
EDIT CALC TESTS
1:1-Var Stats
2:2-Var Stats
3:Med-Med
4▣LinReg(ax+b)
5:QuadReg
6:CubicReg
7↓QuartReg
```

4. Press $\boxed{\text{ENTER}}$ to get back to the home screen. To get the calculator to compute the regression equation and correlation for **L1** and **L2** you need to input **L1,L2** after **1-Var Stats**.[4] By having the lists in the order **L1,L2**, your calculator will make the **L1** list (height) the input or independent variable and the **L2** list (arm span) the output or dependent variable when it calculates the regression equation. To calculate the regression equation press $\boxed{\text{2nd}}\boxed{1}\boxed{,}\boxed{\text{2nd}}\boxed{2}$. Pressing $\boxed{\text{ENTER}}$ will give you the slope and y-intercept for the regression equation and the correlation. The regression equation for this set of data can be written as $y = 0.874575119x + 21.35316111$. The correlation is 0.9043440664.

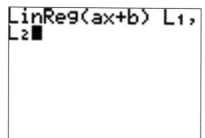
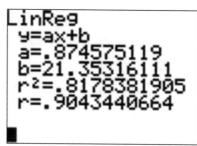

5. To graph the regression line along with a scatterplot of the data, you can type in the equation manually or use a short-cut. The short-cut allows you to automatically have the regression equation stored as a function so you can easily graph it. To do this, go back to the previous step after you have **LinReg(ax+b) L1,L2** on your screen. You

---

[4] The calculators default list is **L1** and **L2**, so if your list of numbers is in **L1** and **L2**, you do not really need to enter **L1,L2**. However if it is in any other lists, you must identify those particular lists.

now need to insert **,Y1** after **LinReg(ax+b) L1,L2**. To do this press $\boxed{,}$ $\boxed{\text{VARS}}$ $\boxed{\triangleright}$ $\boxed{\text{ENTER}}$ $\boxed{\text{ENTER}}$. Your screen should look like the following picture on the left. Now by pressing $\boxed{\text{ENTER}}$, the calculator will calculate the regression equation and store this equation in **Y1**. Graph this along with the scatterplot by following the directions given earlier in this appendix.

 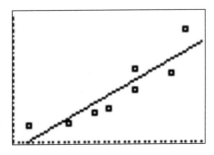

Finding an exponential regression equation

1. Press $\boxed{\text{STAT}}$ then choose **Calc**.

2. Choose **0: ExpReg**.

3. Type **L1,L2** if these are the two lists containing your data. Also include the variable **Y1** so the calculator will store the exponential function as **Y1** (which makes it easier to graph). Do this by pressing $\boxed{\text{VARS}}$ and then choosing **Y-VARS, 1:Function**, then **1:Y1**.

4. Press return. The calculator will give you the values of the y-intercept and growth factor. [Note: The letters for the constants that the calculator uses are opposite those used in the reading.]

5. Graph the scatterplot together with the exponential function given by the calculator.

Finding a power regression equation

1. Press $\boxed{\text{STAT}}$ then choose **Calc**.

2. Choose **A: PwrReg**.

3. Type **L1,L2** if these are the two lists containing your data. Also include the variable **Y1** so the calculator will store the power function as **Y1** (which makes it easier to graph). Do this by pressing $\boxed{\text{VARS}}$ and then choosing **Y-VARS, 1:Function**, then **1:Y1**.

4. Press return. The calculator will give you the values of the coefficient and the exponent for your power function. [Note: The letters for the constants that the calculator uses are different from those used in the reading.]

5. Graph the scatterplot together with the power function given by the calculator.

Index

About the Authors

Janet Andersen has been a member of the Hope College Mathematics Department since 1991, Director of the Pew Midstates Science and Mathematics Consortium since 2002, Chair of the Mathematics Department from 2000 to 2004, GEMS (General Education Mathematics & Science courses) Coordinator from 1996 to 2001, and Director of General Education from 1998 to 2000. She taught high school in East Texas for four years before attending graduate school at the University of Minnesota.

She has been the Principal Investigator for three NSF curriculum grants. The second grant, awarded in 1997, led to the development of a general education mathematics course tied to two general education science courses at Hope College. Her co-author, Todd Swanson (Mathematics) collaborated with her on this project, along with, Ed Hansen (Geological and Environmental Science), and Kathy Winnent-Murray (Biology). The materials from the mathematics course are contained in *Understanding our Quantitative World*. The first NSF grant, awarded in 1993, resulted in *Projects for Precalculus* and *Precalculus: A Study of Functions and Their Applications*. Her third grant, awarded in 2000, resulted in the development of a co-taught mathematical biology course. She also enjoys being with her family, contra dancing, playing Euro board games, and reading mysteries.

Todd Swanson received a BS in mathematics from Grand Valley State University in 1985 and then taught high school mathematics for two years. He received an MA in mathematics from Michigan State University in 1989 (where he received the Excellence in Teaching Award for Senior Graduate Students). He has taught at the college level since 1989 and has been at Hope College since 1995.

His other books, *Projects for Precalculus* (published in 1997 and awarded the Innovative Programs Using Technology Award) and *Precalculus: A Study of Functions and Their Applications* were co-authored with Janet Andersen (Hope College) and Robert Keeley (Calvin College).

Much of Todd Swanson's teaching time at Hope is devoted to Introductory Statistics. He has written numerous laboratories that involve the incorporation of Minitab and are aimed at trying to get students to understand the concepts while exploring real world data. He has also taught liberal arts mathematics, precalculus, calculus, mathematics education courses, and an introduction to writing proofs. Outside of work he can be found working around the house, transporting one of his children to soccer or baseball practice, and participating in some outdoor activity.

Governors State University
Library Hours:
Monday thru Thursday 8:00 to 10:30
Friday 8:00 to 5:00
Saturday 8:30 to 5:00
Sunday 1:00 to 5:00 (Fall
and Winter Trimester Only)

Education Facility
Security Handbook

GOVERNORS STATE UNIVERSITY
UNIVERSITY PARK
IL 60466